How to Raise

How to Raise a Puppy moves away from the traditional approach to raising puppies, focused on obedience and control, and instead takes a holistic, dog-centred approach. Drawing on research into how dogs naturally rear their young, and how dogs have evolved to behave and spend their time, it supports a new way of sharing our lives with our dogs. It also offers advice on dealing with some of the common challenges people experience with puppies, and tips for managing adolescence.

A much-needed resource for dog trainers, veterinarians, and behaviourists to recommend to clients, this book conveys a powerful message to help overcome all-too-common issues many people have with their puppies. Packed with practical advice, it offers an overdue 'puppy perspective', with respect for our dogs as sentient beings at its core.

Dublin-based **Steph Rousseau** is a dog behaviourist and owner of Steph's Dog Training. She is also the Founder of Happy Office Dogs, which offers support and resources to dog-friendly workplaces as well as to those considering introducing dog-friendly policies. Her book *Office Dogs: The Manual* was published in 2019. It has also been published internationally and multilingually. She sat on the Board of the Pet Dog Trainers of Europe from 2017 to 2020 and since 2013 has worked with hundreds of dogs and their humans in both London and Dublin. Rousseau has spoken at various events around the world, including the Pet Dog Trainers of Europe (PDTE), AGMs in Vienna, Austria, and Durham, UK, and the Dog Symposium in Norway. She has appeared on national television and radio in Ireland, as well as on podcasts and radio stations in America, Australia, and beyond. Prior to dedicating herself to all things 'dog', she studied at Trinity College in Dublin and Cambridge University in the UK.

Turid Rugaas is an internationally renowned dog and equine behaviourist; her series of courses for dog and horse instructors is known and respected worldwide. In 1984 she set up her own dog school and conducted several studies on dogs, one lasting two years and resulting in the publication of the book *On Talking Terms with Dogs*, still a bestseller in many countries. Following her attendance at the Human-Animal Bond conference in Montreal, Canada, in 1991, Turid has travelled the world three to four times a month, giving talks and seminars, and educating dog trainers in 17 countries. She has written three books: *Calming Signs: On Talking Terms with Dogs*, *Barking: The Sound of a Language*, and *My Dog Pulls, What Should I Do?* Rugaas earned the Norwegian King Harald's medal in 2018 for her work with dogs and today she runs a new centre – www.hundelandnordvest.com – together with her daughter Linda.

Praise for the Book

"Rousseau and Rugaas's *How to Raise a Puppy* is a gift to dogs and the people who love them. With this book in hand, we can provide puppies the foundation for becoming happy and healthy adults. Armed with decades of combined experience with dogs, the authors not only explain what puppies need but, more importantly, they help us understand why. They challenge some persistent, harmful puppy-raising practices and offer kinder and more effective alternatives. I wish everyone would read this book before making the decision to get a puppy, again after deciding to get a puppy, and then again a few more times after that."

Jessica Pierce, Ph.D., bioethicist, philosopher, and writer.
Author of *Run, Spot, Run: The Ethics of Keeping Pets* and
The Last Walk: Reflections on Our Pets at the End of Their Lives

"Deciding to bring a dog of any age into your home and heart is a huge decision. Puppies, perhaps especially, need all the love they can get, and *How to Raise a Puppy* clearly shows why the only way to teach these deeply feeling sentient beings how to adapt to a human-oriented world is by using force-free, positive methods. This excellent guide should be required reading for everyone who wants to become a trusted guardian for these deeply emotional canines and develop and maintain a mutually respectful, beneficial, and safe friendship."

Marc Bekoff, Ph.D. author of *Dogs Demystified: An A-Z Guide to All Things Canine* (forthcoming) and
coauthor of *A Dog's World: Imagining the Lives of Dogs in a World without Humans*

"This ground-breaking book is the answer for so many puppy owners who want to stay away from training methods that cause stress in their dogs, but have been unaware of the positive alternatives. Until now! *How to Raise a Puppy* is practical, full of dog-friendly advice, examples and background information. It enables the reader to guide their puppy harmoniously into our shared world, while empowering him and acknowledging him as a sentient being. No wonder I read this book with so much warmth and pride, but also relief: 'Yes, it is finally here! The book every pup owner should read.'"

Rachaël Draaisma, MSc, author of *Language Signs and Calming Signals of Horses* and *Scentwork for Horses*

How to Raise a Puppy

A Dog-centric Approach

Stephanie Rousseau
Turid Rugaas

CRC Press
Taylor & Francis Group
Boca Raton London New York

CRC Press is an imprint of the
Taylor & Francis Group, an **informa** business

First edition published 2023
by CRC Press
6000 Broken Sound Parkway NW, Suite 300, Boca Raton, FL 33487-2742

and by CRC Press
2 Park Square, Milton Park, Abingdon, Oxon, OX14 4RN

CRC Press is an imprint of Taylor & Francis Group, LLC

Library of Congress Cataloging-in-Publication Data
Names: Rousseau, Stephanie (Dog trainer), author. | Rugaas, Turid, author.
Title: How to raise a puppy : a dog-centric approach / Stephanie Rousseau, Turid Rugaas.
Description: First edition. | Boca Raton, FL : CRC Press, 2022. | Includes bibliographical references and index.
Identifiers: LCCN 2022017962 (print) | LCCN 2022017963 (ebook) | ISBN 9781032304502 (hbk) | ISBN 9781032304496 (pbk) | ISBN 9781003305156 (ebk)
Subjects: LCSH: Puppies. | Dogs.
Classification: LCC SF427 .R795 2022 (print) | LCC SF427 (ebook) | DDC 636.7/07--dc23/eng/20220713
LC record available at https://lccn.loc.gov/2022017962
LC ebook record available at https://lccn.loc.gov/2022017963

ISBN: 9781032304502 (hbk)
ISBN: 9781032304496 (pbk)
ISBN: 9781003305156 (ebk)

DOI: 10.1201/ 9781003305156

Typeset in Palatino
by Deanta Global Publishing Services, Chennai, India

Contents

Acknowledgements *vii*
Introduction *ix*

1 **PREPARING FOR PUPPY** 1

2 **BRINGING PUPPY HOME** 23

3 **MEETING YOUR PUPPY'S NEEDS** 37

4 **LEARNING LIFE SKILLS** 65

5 **COMMON CONCERNS FOR PUPPY OWNERS** 85

6 **MANAGING RELATIONSHIPS** 103

7 **ACTIVITIES TO TRY WITH YOUR PUPPY** 123

8 **BEYOND PUPPYHOOD – WHAT TO EXPECT NEXT** 133

9 **TURID'S CASE STUDIES** 141

10 **CLOSING REMARKS** 145

An update from Scout – now eight months old 147
References and further reading 149
Index 153

Acknowledgements

Back in 2016, Stephanie was attending Turid's International Dog Trainer's Education. Someone asked Turid if she could recommend a puppy book in English. 'No', came the characteristically blunt response! A number of years later, at a PDTE AGM, no puppy book meeting Turid's approval having yet made an appearance, Stephanie asked Turid if she would consider co-writing one with her. Initially, Turid suggested Stephanie write it, and that she would help. As Stephanie's aim was to create a puppy book that was underpinned by Turid's philosophy, however, she was keen that this should be a joint project. She persevered, and Turid agreed! So firstly, Stephanie would like to thank Turid for agreeing to embark on this project – she's delighted there is now a puppy book on the market that meets with Turid's approval!

Secondly, both authors would like to thank their friends, colleagues, and clients who contributed to this book by way of photographs and case studies. They are particularly grateful to Georgia and Scout who diligently (and honestly!) recorded those early months together, warts and all. Also, to Shane Ó Cathasaigh, Bran, and family who allowed us to document their experiences of bringing a puppy into a busy, young family. Our PDTE colleagues, most of whom have been inspired or trained by Turid, have been very obliging in providing photographs that document the PDTE approach to living with dogs.

Turid would like to extend her thanks to Julia Robertson for sharing so much information about the physical considerations involved in raising a puppy. She is also grateful to her many students who have been involved in her observational studies, looking at behaviours such as sitting, urination, and more. Their work provided insights which had not previously been available and which have influenced Turid's approach to working with dogs.

For her part, Stephanie would like to acknowledge Dr Amber Batson, who, through her courses and seminars, first introduced her to the concept of the canine ethogram and drew her attention to some of the studies, mentioned in the book, highlighting normal canine patterns of behaviour when humans are not involved.

Finally, again on a personal note, Stephanie would like to thank the people who read drafts of the book and offered feedback and suggestions – Adele, Rita, Georgia, Sam, Audrey, and particularly her husband Tom, who patiently and critically appraised numerous drafts!

The authors hope the reader will enjoy the fruits of their labour, and that the dog-centred approach advocated in this book will help make raising your puppy an enjoyable experience for all involved.

Introduction

PUPPYHOOD IN NATURE

In the past, dog training often took its lead from perceived behaviour in wolves. Over the past 20 years, however, it has been generally acknowledged that there are many problems with this model. In short, the studies observing wolf behaviour on which this model was based were themselves flawed and the behaviour misinterpreted.

Read more about the problems with outdated 'dominance'-based theories of dog training:

- James O'Heare: *Dominance Theory in Dogs*
- R. Coppinger and L. Coppinger: *Dogs: A New Understanding of Canine Origin, Behavior and Evolution*
- John Bradshaw: *In Defence of Dogs*

Furthermore, dogs and wolves are different species, albeit with a common ancestor. It is now widely acknowledged that we can learn much more about dog behaviour and what it means to be a dog by watching free-ranging dogs who live with minimal human interference.

As a result, populations of free-ranging dogs in Europe and Asia have increasingly been the subjects of observational studies. We have learned much about how dogs naturally choose to spend their time, what they eat, and, crucially, how they raise their young.

This information can play a significant role in helping us, as their human caretakers, to create an environment where they can grow and flourish, in a manner that is in keeping with their natural essence as dogs.

Read more in these studies on free ranging dogs:

- Pal, Ghosh, and Roy. (1998). Dispersal behaviour of free-ranging dogs (*Canis familiaris*) in relation to age, sex, season and dispersal distance.
- Paul and Bhadra. (2016). Do dogs live in joint families? Understanding allo-parental care in free-ranging dogs.
- Pal. (2005). Parental care in free-ranging dogs, *Canis familiaris.*
- Boitani, Francisci, Ciucci, and Andreoli. (1995). Population biology and ecology of feral dogs in central Italy.

So what does puppyhood look like when humans aren't involved?

- Research on the dispersal patterns of dogs (when they move away from their mothers) showed that two-thirds of puppies never leave their mothers, and for those that do, the mean age of dispersal is about nine months.
- Weaning typically happens at around 7–13 weeks.
- In some cases, the father also remains to help care for the puppies.
- Alloparenting, when other members of the group help care for the young, is common and often provided by the extended female family (aunts, grandmothers etc). Incidentally, alloparenting has also been reported in companion dogs.
- Young puppies are cared for by adult dogs at all times.
- Puppies are generally not left to cry. Their mothers are very responsive and do not wish them to be distressed. Additionally, crying puppies attract predators.
- Puppies will sleep alongside their littermates and their mother.

In contrast, pet dogs are typically separated from their mother and littermates at eight weeks of age and weaned at five to six weeks to facilitate this. They will often not spend any meaningful time with another dog for quite some time after they are brought home. Additionally, we humans generally expect quite high levels of independence from puppies (who would naturally be in receipt of high levels of maternal and family care at this age), leaving them to sleep alone and sometimes to spend many hours alone during the day.

In this book, we will look at what we can do to bring our puppy's lives more in line with the natural order of things. We'll also help you gain the knowledge and understanding you need to raise a happy, relaxed, and confident dog.

It is our view that in order to grow into a great dog, your puppy does not need lots of training, he just needs love, understanding, some good habits, and the chance to be a dog!

An Indian street-dog caring for her puppies while the puppies' grandmother stands guard. Courtesy Preeti Sushama.

Grandmother babysits while the mother dog is away. Courtesy Preeti Sushama.

INTRODUCING SCOUT

Throughout this book, we'll check in with Scout, whose mum Georgia shares her experiences of surviving those first few months with a new puppy in the house!

Scout was adopted from Dogs Trust Ireland in 2020 at 12 weeks of age, and joined an adult-only home with a resident dog and cat in situ. He is a collie and the family are experienced with the breed.

Georgia experienced many moments of stress and frustration and at the time of writing is now through the puppy stage.

We hope you enjoy these bites of reality from a real-life puppy family!

Scout, the border collie puppy making himself at home! Courtesy Georgia O'Shea.

1 *Preparing for puppy*

THINGS TO THINK ABOUT

The benefits of having a puppy are many! Getting a puppy is a great opportunity to raise a dog from scratch, to watch them grow and develop into a wonderful friend and family member. Puppies are also incredibly cute and can provide hours of entertainment. You may even love them more than you've ever loved another human being!

But it's also totally normal to have moments of despair, stress, and worry. A puppy is a huge commitment, and most people find puppies pretty hard work. It is common for them to drive people to tears and cause high levels of frustration.

The struggles people most often complain about are sleep deprivation, constant cleaning up after toileting accidents, chewed hands, and destroyed possessions. Forewarned is forearmed, however – thinking about how you're going to deal with these things in advance can help take some of the stress away when you're living through it, and we will do our best to prepare you.

It is wise, however, to give some consideration to the following before deciding to embark on this journey with a new puppy:

- How will you make sure your puppy will have constant care for at least the first four to five months of his life?
- How will your current household, especially children, take to the new family member?
- Are you committed to the dog that the puppy will grow into as he leaves the cute puppy phase?
- Who will mind your puppy when you go away?
- What dog-care options are planned for when you need to be out of the house in the future? For an adult dog, four to five hours is ideally the maximum amount of time they should be left without at least a comfort break. For longer absences, think about a dog walker, or even having a neighbour drop in to say hello and let them out to toilet.
- Will you get pet insurance? If not, how will you cope with any big vet bills that come your way? Surgery after an accident, or an ongoing illness, can cost thousands of euro. Some people, rather than paying for insurance, choose to put away a set amount of money each month to deal with any unexpected health costs. Bear in mind that some breeds are more prone to health conditions that may require expensive interventions.
- If you're not getting health insurance, what about liability insurance? In the UK, for instance, if your dog escapes and causes an accident, you are personally liable

DOI: 10.1201/9781003305156-1

for that accident. This sort of coverage may be available through your house insurance or may need to be purchased as separate pet insurance.

- Are you prepared for the damage your puppy will most likely do to your house? Households with pristine cream carpets and beautiful antique furniture, take heed!
- Have you calculated the likely costs your puppy will incur in the short, medium, and long terms? Vet fees, insurance, dog walkers, holiday care, food, and equipment all add up. It has been estimated that with an average lifespan of 10 to 15 years, a dog will cost you about €2000 to €2500 a year (although this seems conservative to us!). Bigger dogs will cost more than smaller dogs and the initial cost that you pay the breeder really is the tip of the iceberg!

An Australian labradoodle puppy surveying his surroundings. Courtesy Jolly Doodles.

For many people, the joy and companionship your dog will bring for the next 10 to 15 years more than compensates for any of these potential issues. So, once you're happy you've considered these key things, it's time to get cracking!

UNDERSTANDING DOG BREEDS

A dog's breed is only one of the factors that will impact their personality and behaviour. There are certain characteristics which are simply 'dog' characteristics, shared by all breeds, and some characteristics which may be more prevalent in specific breeds. On top of this, your dog's individual personality and the environment in which they live will strongly influence who they grow up to be.

The breed you choose still deserves due consideration, however. When it comes to what breed of dog to choose, we are often led by our eyes. Fluffy poodle crosses have become very popular in recent years, as have breeds with shortened snouts (brachycephalic breeds). However, choosing the breed you find most aesthetically pleasing is not always a recipe for success! Different breeds of dogs will often have different temperaments and inclinations depending on the type of job they were originally selected to do.

EXTREME FEATURES

Dogs that have been bred to have extreme features – very short snouts, very short legs, very long bodies, extra-large body mass, an unusual coat – will often have physical problems with that feature as a result.

Brachycephalic breeds (those with shortened heads such as French bulldogs, pugs, boxers, and Pekingese) will often have problems breathing and may need costly surgeries to open their airways. Long-backed dogs with short legs, such as dachshunds and basset-hounds, are more susceptible to back problems. The coats of poodle crosses are prone to matting and can require a great deal of maintenance.

Often the more 'normal' a dog looks, the better they will function from a physical perspective, so this is another thing to bear in mind when choosing a breed.

A French bulldog puppy showing the shortened snout of a brachycephalic breed. Courtesy Louise Haughton.

INHERENT BEHAVIOURS SHARED BY ALL BREEDS

Before selecting a particular breed, it is important to remember that all dogs, regardless of breed, have evolved over millions of years (long before they crossed paths with humans) to carry out certain behaviours. We cannot erase these tendencies, nor should we try. But recognising these behaviours when they occur and knowing where they might have come from, can help us to understand them better and to respond appropriately.

HUNTING

Our dogs' hunting instincts mean they are prone to chasing things that move – including cars, children, joggers, cats, birds, and so on. All dogs have inherited the motor patterns to carry out this behaviour and this inclination will always be there to some extent. Indeed, assisting humans with hunting was one of the earliest collaborations between man and dog. It is a natural, automatic reaction when something moves away, but we can learn to manage it in daily life.

Is your dog using his hunting instincts more than you would like? Try the following:

- Manage the situation by not putting your dog in situations where he is likely to chase, or by keeping him on a lead when this is unavoidable.
- Cut out other activities that use the same motor pattern (i.e. if you want your dog to stop chasing squirrels, do not throw balls or otherwise encourage him to practise that prey-chase response; equally, if you do not want him to chase cars, do not allow him to chase squirrels either!).
- After your dog has had several weeks of not practising the behaviour, you can start to reintroduce the trigger at a low and controlled level; keeping your dog on a lead and harness initially, work on using your distraction noise (page 70), to take your dog's focus away from whatever they want to chase.

SCAVENGING

Scavenging can be a more energy-efficient and reliable way of finding food than hunting and is another hardwired behaviour in dogs. Amongst free-ranging dogs, scavenging behaviour is much more prevalent than hunting behaviour. For a dog, scavenging behaviour can also manifest as searching or tracking animals, people, or objects, with their noses. Searching is one of a dog's strongest instincts, and we must try to allow them to do so as much as possible – it is in their genes!

Following her nose! A papillon puppy does some scent trailing.

SELF-DEFENCE

For dogs, being able to defend themselves is necessary for their own survival, as well as that of their family, and their species. All animals, human and non-human, defend themselves when they feel threatened. In dogs, self-defence will often manifest as growling, air-snapping, or even biting. Dogs are inherently peaceful, however, and these behaviours will usually only be used when other methods (such as calming signals, page 103) have been tried and have failed. Alarming as this may be to us, it is a normal, natural thing to do, and not something that should be punished. When humans act in self-defence, it is recognised as a normal response to a threat. However, when dogs do the same, they are (unfairly, we think) often punished and labelled as aggressive.

COMMUNICATION

As social animals, often living in a family or social group, dogs have strong bonds within their social circle. They also need to be able to communicate with each other in order to avoid problems or conflict, and they use these same communication skills with us. Problems can arise if we do not recognise, or misinterpret, their attempts to communicate with us, so we need to be open to learning how they communicate (see more on page 103).

BREED SELECTION

From the time when dogs first started to share their lives with humans, we have been taking advantage of these traits and breeding selectively to strengthen or weaken them.

For instance, if one family needed a strong guarding dog, they would breed their dog with another one who seemed like a good guarder (strong self-defence tendencies). These offspring inherited some of those guarding skills, and when the same process was repeated for that generation, these traits became even stronger. If another farmer needed dogs for hunting, and he hunted big game, he would look for dogs with qualities suitable for that activity (hunting behaviour). Breed selection had started.

Today, we have hundreds of breeds that have been raised for different purposes – different hunting methods, many different ways of herding and guarding – but they have all been bred for something. It is likely your dog will show you some of the skills they were bred for in daily life, and it should not come as a surprise when they do!

Historically, different types of dogs were simply known by the sort of work they used to do – herding dogs, guarding dogs, hunting dogs. It was during the Victorian period that fanciers began to create breeds that looked a certain way, or possessed certain skills. However, all of the breeds we have today can be categorised into certain groups, which, again, reflect the jobs they were bred to do. Dogs within these groups will often share certain key features and behavioural tendencies.

LOCAL BREED TRADITIONS

The people of different countries and even districts often have their own hunting traditions and many have a dog specially bred to help them catch whatever bird or species they hunt.

Each country in Europe has many local breeds that started this way – some of which have never made their way onto any official list of breeds, but are very well known in their own country, even if there are relatively few of them.

Bred to retrieve birds, many retrievers love having things in their mouths and will happily carry a toy, some groceries, or anything else they can find. Courtesy Siobhán Kavanagh.

Table 1.1 Breed Groups and Characteristics

Breed group	Historical breeding purpose	Key features	Advantages	Things to be aware of/ potential challenges
Sighthounds Examples include: • Greyhound • Whippet • Deerhound • Afghan hound • Ridgeback	• Hunting.	• Fast-moving hunters who chase prey on sight. • Able to run at speed but not for long distances. • Deep chests for increased intake of oxygen. • Long thighs for explosive power but not designed for sitting, and long necks for balance. • Not bred for guarding skills.	• Limited barking unless very stressed. • Only require short bursts of exercise to be happy. • Tend to spend large amounts of time relaxing.	• Not built for a lot of walking. • Need to have regular opportunities for a short, fast run. • Quite independent-minded. • They can become easily excited and can be prone to chasing moving objects e.g. joggers and bikes. • To manage this urge to chase, be cautious and take away the opportunity for them to practise this behaviour.
Scent hounds Examples include: • Beagle • Foxhound • Basset hound • Dachshund	• Hunting.	• Bred to pick up the scent of the prey and follow it. • They bark as they go to let the hunter know where they are – this communication is necessary for the hunt. • Often will have long ears, which may be helpful in collecting the scent and keeping it close to the nose. • Tend to possess large nasal cavities.	• Usually pleasant-natured and loyal, with obliging temperaments.	• Often quite vocal dogs, with a tendency to bark, howl, or bay. • Providing them with plenty of opportunities to follow their noses will be important for their well-being.

(*Continued*)

Table 1.1 (Continued) Breed Groups and Characteristics

Breed group	Historical breeding purpose	Key features	Advantages	Things to be aware of/ potential challenges
Herding/Pastoral Examples include: • Collie • Mountain dog • Rottweiler • Kelpie • Corgi	• Bred to herd domestic animals. • Different breeds specialise in different types of herding.	• Come in many shapes and sizes, depending on the work they were bred to do. • Lighter breeds like collies are built for fast movement, herding sheep. • Heavier herding breeds like the Dogue de Bordeaux are built for stability. They move cattle or sheep or pull carts. • Kelpies and Australian cattle dogs are specialised to work with huge flocks.	• This varies depending on the sort of work they were bred to do. • Collies are quick to learn, and great at caring for their families. • Mountain dogs work more slowly and take their jobs very seriously. • Kelpies and Australian cattle dogs work in groups so tend to get on well with other dogs.	• These are guarding breeds and can be quick to guard their families and properties. • Collies need to use their brains and have a proper job. They can start to herd anything in the household that they think needs managing – including children, chickens, and rabbits. • Corgis can use their working method (pinching ankles) on humans too!
Terriers Examples include: • Jack Russell terrier • Border terrier • Cairn terrier • Bull terrier	• Bred for hunting vermin, often underground. • Bull terriers were bred for blood sports like fighting and baiting.	• Often small, to allow them to go underground after the prey. • Bull-terriers are bulkier than their working terrier counterparts, but can share their stubborn streak.	• Terriers are lively, clever, and affectionate. • Bull terriers are known for being affectionate with humans and very joyful.	• Terriers tend to be independent, determined and tenacious – they do not give up easily. • Bull terriers require early socialisation around other dogs to ensure better social skills with other dogs as they get older.

(*Continued*)

Table 1.1 (Continued) Breed Groups and Characteristics

Breed group	Historical breeding purpose	Key features	Advantages	Things to be aware of/ potential challenges
Gun dog Examples include: • Retrievers (Labrador, golden retriever) • Pointing dogs (pointer, setter) • Flushing dogs (spaniel)	• Assisting hunters by finding and retrieving game.	• Increasingly, differences can be seen between working and show strains of these breeds. • In personality, gun dogs bred to work will be more driven and focused than their show counterparts. • Physically, show dogs may be bred to have more exaggerated features such as overly long ears, which would be a hindrance to a working dog, or to be bigger and less lean.	• As they are bred to work with both people and other dogs, they are usually well disposed to both. • Will enjoy having a job to do that involves interacting with their family.	• Retrievers often like to carry something in their mouth whenever they are excited and can be difficult to keep out of water. • They can be overly exuberant and friendly around other dogs if they do not learn appropriate social skills.
Utility dog Examples include: • Poodle • Dalmatian • Schnauzer • Bulldog • Spitz	• These are dogs that were originally bred for a purpose, but no longer serve that purpose. • Poodles were bred to retrieve waterfowl, for instance. Dalmatians were bred to be all-rounders: guarding the camp, running with chariots when they moved. They have also been used for poaching – moving around silently.	• Incredibly varied.		

(*Continued*)

Table 1.1 (Continued) Breed Groups and Characteristics

Breed group	Historical breeding purpose	Key features	Advantages	Things to be aware of/ potential challenges
Toy dog Examples include: • Pug • Pekingese • Cavalier • Bichon frise	• Small dogs bred for companionship.			

PAWS FOR THOUGHT: BREED RESEARCH

Do some research into the breed of dog you're thinking of getting. Consider:

- What job was the breed meant to do?
- How do they do that job?
- Were they bred to work alone or with people?
- How will this impact their temperament?
- Are they prone to particular health issues?
- What sort of coat do they have and what sort of maintenance is required?

CHOOSING THE RIGHT PUPPY

One of the first things you can do not only to increase your chances of having a healthy, happy puppy, but also to improve the welfare of puppies more generally, is to pick a good breeder. Sadly, many puppies are still handed over in car parks, or sold from homes that are not actually where the puppies were born. These puppies have usually been born in deplorable conditions on puppy farms.

Spending time researching breeders and their practices is invaluable. Some things to look out for in a reputable breeder include the following:

- The breeder has nearly as many questions for you as you have for them! A good breeder who cares about their dogs will want to know that their puppies are going to a good home.
- They have a policy of taking the puppy/dog back if you ever find yourself in a situation where you can no longer keep them or if the puppy has a hereditary disease.
- They are happy to let you meet the mum (dam) and to see her with the puppies.
- They allow for multiple visits so the puppy can become familiar with you.
- The dam looks happy, relaxed, and well cared for.
- The puppies have been gradually and gently exposed to the sorts of things they may encounter in the world.
- They can provide convincing references for previous puppies.
- They are registered with an appropriate body, for example, your country's Kennel Club.
- They know about the breed and offer sound advice about caring for your puppy.
- They have veterinary records showing that the puppies have been wormed and vaccinated where appropriate.

Mum keeping a watchful eye on her litter of puppies. Courtesy Myriam Deckers.

Warning signs that something might be off:

- The breeder has many different breeds of dog for sale.
- The puppies or the dam look unwell, unhappy, or uncared for.
- They always have puppies for sale.
- They make excuses as to why you cannot see the mother and the pups together or as to why you can't visit.
- The dam is there but is not interacting with the puppies. This can be a sign of a 'fake mum'.
- They are keen for you to take the puppy home as quickly as possible.
- The dam was bred before she was two years of age or after about six years of age.

Understandably, once you have met a puppy, it is very hard to walk away without him. This is particularly true if you feel that you are leaving a puppy in poor condition or on a puppy farm. However, it is important to remember that you are simply supporting the trade in buying a puppy from a breeder like this. Ending up with a puppy from a puppy farm increases demand for puppies bred in these conditions and sentences female dogs to a lifetime of having litter after litter, each one taken away from them.

Furthermore, you could be creating a rod for your own back! Some undesirable characteristics, such as fearfulness, are inherited. An unscrupulous breeder will not be breeding for character. Mother dogs who are forced to have litter after litter of puppy in poor conditions, and then constantly have the puppies removed, will be very stressed, and this can have a negative effect on the puppies. Studies have also shown an increased incidence of behavioural disorders such as fear and aggression towards other dogs and humans in commercially bred dogs.

Doing as much research as possible into the breeder before going to meet the puppy can help reduce the chance of you ending up in this difficult situation.

If you suspect that you have come across a puppy farm, you may be able to report the seller, depending on the laws in your jurisdiction.

WHICH PUPPY TO CHOOSE?

People often want to know which is the best puppy to choose from a litter. It is often recommended that you should opt for a confident, curious puppy. However, at such a young age, it is almost impossible to predict how a puppy will turn out. Their personalities can change from week to week – and often even more quickly than this. The environment they grow up in will have a more significant impact on the dog they grow up to be. For instance, a dog could have a latent tendency to be nervous and unsure, but if we provide an environment in which they learn to cope, these tendencies might never come forward. Dogs are survivors and will easily adapt to a new environment if given a proper chance to do so.

ADOPTING A PUPPY FROM A SHELTER

Puppies are often available to adopt from shelters, when they are found abandoned or when a pregnant bitch is taken into a rescue. These dogs are often mixed breeds, and can be a great option for somebody who would like a puppy but does not mind the breed of the dog.

However, you may meet some extra challenges rearing these dogs.

In situations where puppies were abandoned, they may have already had some traumatising formative experiences around attachment and resources. It is likely that their mother was not loved and cared for during her pregnancy, and the puppies may have been subjected to high levels of stress hormones in the womb, which may affect how the dogs react to stressful situations for the rest of their lives.

These puppies may also have been taken away from their mothers at a very young age. It has been reported that puppies who are separated from their mother and littermates before eight weeks tend to bite harder and more readily than those separated after that point, and may develop other problems later in life, such as inter-dog issues, attachment-related problems, anxiety, and reactivity. Of course, this is also a risk with unscrupulous breeders and puppy-farm puppies.

Puppies born in rescues may also face challenges arising from their early experiences. Although many shelters make great efforts to provide the best environments that they can for their dogs, these can still be stressful places, and not the ideal environment for pregnant or nursing dogs.

Additionally, these pups are unlikely to have been bred for temperament the way that a registered breeder will, hopefully, have bred their dogs. Therefore, you may be more likely to end up with a dog who has inherited fearful tendencies, or other less desirable attributes.

On the other hand, many of these puppies make fantastic pets with no additional challenges, and it is certainly arguable that in a world with so many dogs in need of homes, rescuing a puppy from a shelter is a more ethical decision than buying one from a breeder. Additionally, love, patience, and the right information can help overcome many of the issues that may arise.

Scout, despite a less than ideal start, is proving to be a charming puppy! Courtesy Georgia O'Shea.

WHAT TO BUY BEFORE PUPPY ARRIVES

Everybody wants to be prepared for their new canine arrival, but with so many items available for dogs and puppies it can be hard to know where to start! We've compiled a list of the things you need, (and the things to avoid) to set you on the right path.

COSY BEDS

As we'll discuss later on, puppies need a lot of sleep! Providing them with lots of comfortable places to sleep can help them achieve the shut-eye they need. You can find more information on the things to consider when buying your puppy a bed in the Sleep section of this book. It's worth investing in decent beds, however, as your puppy will hopefully spend a lot of time sleeping!

A comfortable bed can help your puppy get the sleep they need. Courtesy Georgia O'Shea.

✓ NEWSPAPERS OR PUPPY PADS FOR TOILETING ACCIDENTS

Be prepared for accidents… Newspapers may trump puppy pads for a number of reasons. Firstly, it is often said that dogs can have a life-long preference for going to the toilet on the sort of surface they were initially trained on. Soft absorbent surfaces like rugs, beds etc. may be tempting alternatives for dogs trained on soft, absorbent puppy pads! Picking up a bunch of free newspapers, or collecting used papers from friends and family is a cheaper, plastic-free alternative!

✓ A GOOD CLEANING PRODUCT

Enzyme-removing cleaning products are required to fully remove the smell of urine so that your dog can no longer smell it. If the smell is not fully removed, the dog will most likely return to this spot for future toileting business! These enzyme-removing cleaning products can be purchased online or in pet stores. You can even create your own, using common household products such as biological detergents, vinegar, and baking soda. Products that contain ammonia can encourage repeat offences!

SIMPLE HOME SOLUTION FOR PEED-ON CARPETS!

Soak up any pee that you can, using paper towels or newspaper. Stand on the paper to really soak up as much as possible. Then use a mix of equal parts water and white vinegar. Use a spray container to fully wet the area. Soak up any excess moisture again. Now sprinkle with baking powder and rub in with a brush or sponge. Leave overnight or until dry and then vacuum!

The vinegar will neutralise the urine odour and the baking powder will help get rid of the vinegar smell and any residual urine smell.

For hard floors, simply dampen a cloth, pour some white vinegar over it, wring it out, and mop the areas.

✓ DOG BAGS

Compostable, bio-degradable bags are now widely available. Tie a few to your lead so you're never caught short.

✓ FOOD

Your puppy will have enough change to contend with in their first few weeks, so find out what they are being fed by the breeder or shelter, and stick with this for the first while. There will be plenty of time to change to the food of your choice later. Breeders will often provide a supply of the food the puppy is being fed when they are going home in any case.

✓ FOOD AND WATER BOWLS

Bear in mind that puppies will sometimes knock lighter bowls over or pick them up. Dogs will also sometimes object to the taste or smell of plastic or metal bowls, and can be alarmed if their name tag knocks against metal or ceramic bowls when they're eating. You may wish to give your puppy a choice of vessels that you have at home to find out what works best before spending money on dog bowls!

Ensure that the bowl is placed on a non-slip surface so that neither the bowl nor the puppy's paws are slipping as they try to eat and drink. You can buy custom-made mats for this or improvise with towels, bath mats, tea-towels etc.

FOOD-BASED CHEWS

Chewing is a normal, necessary dog activity and, throughout your dog's life, he will need access to things to chew. Opt for natural, untreated animal parts for a long-lasting, relaxing chew. You'll be very glad of the 15 minutes of peace this can buy you!

Natural chews, like this cow's ear, are ideal. Courtesy Federica Iacozzilli.

CHEWS

Try natural, unprocessed chews. Popular items in our house include:

- Pizzles.
- Moonbones.
- Dried ostrich tendons.
- Cows' ears.
- Yakers.
- Raw bones (cooked ones can splinter and cause blockages).

Bear in mind that puppies should be supervised whilst chewing and any pieces which could be swallowed whole should be removed.

NAME TAG AND COLLAR

We recommend light, soft collars for puppies – for smaller dogs, cat collars are often ideal! The collar is only really required to hold the name tag, as we do not recommend attaching leads to collars. Bearing in mind our dogs' superior and sensitive hearing, and the proximity of the name tag to the ears, consider silent name tags. These either attach to the collar like a band or come with non-metal attachments.

HARNESS

A well-fitting harness can save your puppy from damage to their neck and spine when they are out walking. A well-fitting harness should:

- Be Y-shaped at the front.
- Leave the neck clear and sit across the sternum instead.
- Have a rear attachment for the lead.
- Leave space behind the front legs so that it doesn't chafe when the dog walks.
- Allow the shoulders to move freely.

A good harness keeps the sensitive neck area free of pressure, without hindering the dog's natural movement or chafing behind the front legs.

Avoid any harnesses (or other device!) which claim to be 'anti-pull'. Without exception, these work by causing pain or discomfort when the dog does pull, or by pulling their body out of alignment, which will also lead to pain and discomfort.

LEAD

The lead you choose can have a great impact on how your puppy feels. Research from France has shown a marked difference in the heart rate of dogs on long leads as opposed to short leads.

Dogs on longer leads (5m as opposed to 1.5m) had lower heart rates, which suggests that they're more relaxed and spent significantly more time sniffing.

For city dwellers, you can get 3m adjustable leads which can clip onto themselves to reduce to 1.5m for when you are on a small footpath. Longer leads should also be in your arsenal for when you're in open spaces and away from potential danger.

A long lead offering puppy Bran enough space to investigate his environment on his walks.

THINGS TO AVOID

CRATES

We would *strongly* advise against the use of crates. No animal likes to be locked in a cage, and no amount of 'crate training' makes this okay. Crates inhibit natural behaviours, create frustration and boredom, and can have an adverse effect on a puppy's behaviour. They also deny puppies the one thing they need most – sustained social contact.

Be aware of the language used around crates and crating. Crates are often described as a 'den' for our dogs. However, dogs are not, in fact, den-dwelling creatures beyond puppyhood – and puppy dens never have a door! Additionally, a safe place is one where there is free entry and egress. Once locked in a cage a puppy is not safe, they're trapped.

Research has also shown that rather than reduce anxiety, restraining puppies (alone or with littermates) increases reactivity and can have an adverse effect on the social behaviour of an otherwise well-socialised pup.

SOCIAL PAIN

Research has shown that social pain – the feeling that social animals (including dogs) experience when they perceive a weakening or loss of a valued social connection, including physical or psychological separation from their social group – uses some of the same neural mechanisms as physical pain. In fact, it is possible that social pain is actually more distressing than physical pain.

Physically separating them from you in a crate is likely to trigger social pain.

CHOKE/CHECK/PRONG COLLARS

These are designed to inflict pain on dogs if they pull or if the lead is jerked. As well as causing immediate pain, they can also cause long-term damage to the sensitive anatomy of the neck area and cause damage to the eyes from increased pressure. There is no way to use this sort of equipment that does not cause pain.

Slip leads, martingale collars, and half-check chains should be avoided for the same reason.

SHOCK COLLARS

Also known as an e-collar, this is another piece of equipment designed to cause pain, by delivering an electric shock to the dog if he does not do as he's told. Besides being unethical, recent research into the matter 'refute[s] the suggestion that training with an E-collar is either more efficient [than positive reinforcement] or results in less disobedience, even in the hands of experienced trainers' (China et al. 2020). You can read more about the impact of punishment more generally on behaviour on pages 78–80.

☒ HEAD HALTERS

Head collars, which go around the dog's face and attach to their collar, are another 'anti-pull' device. The bridge of the snout, where the head collar will sit, is an extremely sensitive area on a dog, and if the puppy or dog reaches the end of the lead with any enthusiasm (upon seeing a cat, another dog, when startled or excited), their neck will be sharply yanked sideways and up, and, over time, this will lead to soft-tissue damage. As well as the pain they cause to the neck and snout, they can also ride up and rub against the eyes. Additionally, it becomes impossible for the dog to sniff or use their head to communicate (i.e. by turning away).

☒ EXTENDABLE LEADS

Extendable leads can wrap around the legs of dogs, children, or adults, and can potentially cause serious injuries. They also make it more difficult to teach your puppy to walk on a loose lead, as the fact that there is always tension means the lead never feels loose for the dog. Additionally, if you drop the lead (which we all do at some point!), the handle will clatter to the floor, possibly terrifying your dog and causing them to bolt.

PAWS FOR THOUGHT: KEEP A DIARY

When you're buying the various bits and pieces for your dog, we recommend you also buy a diary for yourself!

In this, we suggest keeping a short, daily note of your puppy's activities, diet, behaviour etc. This can be an invaluable resource to monitor both your dog's health and his state of mind. If you realise that your dog is having days when he is very restless, you can look over your diary and see if it coincides with any particular activity. If he develops an unwanted behaviour, you may be able to identify the starting point, or even the trigger. If he has an upset stomach, you can review what he's eaten recently, or perhaps identify that the previous day was just very exciting and that this may have affected his digestive system.

In terms of behaviour, the response to something is not always instant. So, you might find that if you bring your puppy to the dog park in the morning, he struggles to settle that night. Or if there are visitors and a lot of excitement one day, he is grumpy and more nippy than usual the next day. If you know this, you can always ensure that he has plenty of time to recover, or that you cut the offending activity out.

PREPARING YOUR HOME

To save yourself the stress of constantly extracting precious items from your puppy's mouth those first few weeks, take an afternoon to audit the areas to which the puppy will have access. Remember that your puppy will investigate the world with his mouth, so if there is anything that you would rather didn't end up in the puppy's mouth, remove it now! Consider the following items which are common puppy favourites, and how you might keep them safe from your puppy:

- Shoes and socks.
- Soft toys.
- Anything with tassels.
- Wires.
- Plants (check any that you are leaving within reach are not poisonous).
- Anything you particularly love (puppies have a knack for finding these things!).
- Remote controls.
- Chair legs (you may wish to remove any antique or irreplaceable items to the attic).
- Books.
- Anything small they can fit in their mouths.

Puppy Marley feels this boot is an ideal chew toy! Courtesy Laura Allen.

You may also want to block off anything that could pose a risk to the puppy, such as open fires, stoves, or areas where there are a lot of wires that need to remain. If you have already bought a crate, we do not recommend that you use it in the conventional way. However, many crates can be opened up and used to block off unsafe areas, so you can use your crate to keep your puppy out of certain areas rather than to keep them locked in the crate.

Skirting boards and door frames are other common victims of puppy chewing. You may be happy to risk any damage to these items, and if so, that's great! Otherwise, you can try putting some lemon juice or bitter apple spray on them so that they do not smell appealing for chewing. And of course, most importantly, provide a variety of more appropriate items for chewing.

TOILETING PRECAUTIONS

Accidents are inevitable during the first weeks and months. It would be wise to remove rugs and any other items that you really don't want being peed on, and put them back into place when the puppy is older and not peeing everywhere!

SLIPPERY FLOORS, STAIRS AND SOFAS

Although they may seem like bendy, wriggly, indestructible little balls of energy, puppy bodies are actually very vulnerable to damage. The growth plates do not close for many months – as late as 18 months in some breeds, and muscles have not fully developed to support the integrity of the joints. This is the reason that exercise in puppies needs to be carefully moderated. However, over-exercise is not the only risk to your puppy's body. Slipping and sliding on floors can do your puppy untold damage, that will most likely not become apparent until five or six years down the line in the form of arthritis and other musculoskeletal issues.

If you can, put down rugs or another form of non-slip covering on slippery tiles, wood, and laminate flooring.

Similarly, repetitive impact from going up, and particularly down, stairs, jumping on and off sofas and beds, and in and out of cars will put enormous strain on a puppy's body. This sort of impact causes stress on the neck and 'shoulders' as well as on the spinal vertebrae. You may be able to lift your puppy up and down where possible to minimise this impact, use ramps for the car (and even the sofa!), and place cushions or dog beds beside the sofa or bed to absorb some of the impact if the puppy does jump down.

A ramp to help your puppy on to and off higher surfaces can help protect their bodies from undue wear and tear. Courtesy Marina Gates Flemming.

This is particularly significant for long-bodied dogs such as dachshunds, who are at an increased risk of back problems, and for the extra-large breeds such as wolfhounds, mastiffs, and Great Danes.

TAKE-HOME POINTS FROM CHAPTER 1

- Research your chosen breed before making a final decision – choose a breed whose tendencies won't be problematic for you.
- Get stocked up with all of the things you need in advance.
- Prepare your house for your puppy – put away precious items and think about how you can make the home environment more comfortable for your puppy.

2 *Bringing puppy home*

WHEN TO TAKE PUPPY HOME

The typical age for bringing home a new puppy is eight weeks old. Your puppy obviously can't leave their mum until they're fully weaned. In order to have the puppies ready to go home at eight weeks, this usually means weaning at five to six weeks old. As mentioned in the introduction, weaning in free-ranging dogs takes place between seven and thirteen weeks of age.

Your puppy's mum and siblings have much to teach him. Courtesy Elena Barbini.

DOI: 10.1201/9781003305156-2

Finding a breeder who is willing to keep the pups with their mum for an extra two to four weeks, and to wean that bit later, brings with it many benefits. A puppy's mum and siblings have so much to teach him about canine communication, play, bite inhibition, and toilet training, that allowing her to do so can also make your life a lot easier in the long run.

Additionally, taking your pup home at ten to twelve weeks allows him to experience the fear period which occurs at eight weeks, in the safety of his family group, with his mum around to support him.

A TRICK OF TURID'S

If your breeder is determined that you take your puppy home at eight weeks, there is no need to get into a debate. Simply wait until you have all of your plans in place to pick up your puppy, and then tell your breeder that you suddenly need to go away for a fortnight, and can you pick your puppy up on your return!

A mother's love! Allowing a puppy to experience his first fear period in the company of his mum can make it a lot easier for him. Courtesy Elena Barbini.

REMEMBER

In the wild, your puppy would most likely have stayed with his mum for at least nine months. When you take him home at eight, ten, or even twelve weeks, he is still a tiny infant in need of constant care. It's important to realise that the puppy is losing his natural family, but you can help mitigate this by providing him with the same sort of sustained social contact that he would have been receiving from his mother, littermates, and family.

MAKING THE SEPARATION EASIER

Leaving his mum, siblings, and everything he's known in his short life is going to be a huge change for your puppy. Think about what you can do to ease the transition. Can you make regular visits to see your puppy while he is still with the breeder and his mum? Getting to know the person with whom he'll be leaving can help make his removal from his original home a little less frightening.

Can you leave him pieces of your clothing so he gets to know your smell? Smell is one of the dog's most important senses. Sleeping in a T-shirt for a couple of days before you hand it over will ensure it has a nice strong 'you' smell! You could bring items with the smells of any resident pets on them too, and you can also do this smell exchange in reverse – when you meet your puppy for the first time, bring along a blanket that you can leave with his mum. You can then take this home along with your puppy when you pick him up.

FEAR PERIODS

Puppies experience a number of fear periods as they grow. These are times when the puppy is particularly sensitive to frightening experiences. Scary experiences during these fear periods can have a profound and life-long impact on your puppy's behaviour.

The first fear period happens at eight weeks, which is one of the reasons this is not an ideal time to take your puppy home. If you can protect them from the upheaval of being separated from their mother at this exact time, all the better!

Puppies experience a second fear period at six to fourteen months, and possibly a third in early adulthood.

As with the first fear period, the best thing to do during these subsequent fear periods is to avoid anything traumatic, scary, or overwhelming. Treat any fear-based reactions such as barking or lunging with compassion, allow your dog to move away from anything that frightens them, and engage them in nice, calm, confidence-building activities.

INTRODUCING PUPPY TO THEIR NEW LIFE

When the day finally arrives to bring your puppy home, the challenge is to make the transition as stress-free as possible for your puppy, and indeed, for you!

The first step is to get prepared for the journey. It is common for puppies to suffer from travel sickness, as the balance system in the inner ear is not developed at birth, and takes time to develop properly. Toileting accidents are also a strong possibility, so bring towels and seat covers! Check the rules in your jurisdiction about travelling with a dog, and if at all possible, avoid isolating them in a crate for the journey. If permissible, having the puppy sitting on someone's lap in the front passenger seat where they can see out the window can help reduce feelings of travel sickness. Some swear by a quarter of a ginger biscuit before the journey to help settle the stomach.

Remember that even if a young puppy does not actually vomit in the car, it is still likely that they will feel unwell, so it is a good idea to try and restrict long car rides at this time.

Additionally, the excitement of the journey and these new experiences are likely to make your puppy need to urinate more frequently, so be sure to give him the chance to go before he gets in the car. If it is a very long journey, think about whether there is somewhere safe that you can stop every hour or so for the puppy to relieve himself.

Having something suitable for your puppy to chew on the journey can also prove to be a calming distraction for them.

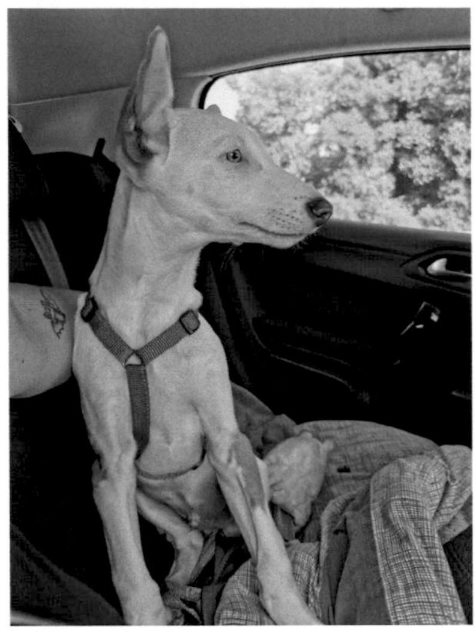

Being able to see out the front window can help alleviate feelings of travel sickness in a puppy. Courtesy Alja Willenpart.

SCOUT'S DIARY – DAY 1

'We pick up Scout in the morning time so as to have as much time with him as possible before facing bedtime. We brought a blanket with our other dog's scent on it to introduce to him in the car. He seems to be a happy, friendly, and engaged puppy that looks for our affection. It's his first time travelling in a car so he is sick on the journey, we all know he must be wondering where his life has gone – what a scary time for him!'

Scout about to make the journey to his forever home. Courtesy Georgia O'Shea.

ARRIVING HOME

First of all, allow your puppy to find his feet in his new environment without interference. He will usually find the water bowl, the food bowl, and a comfortable spot to lie down pretty quickly and without any input. Then, bring him to the spot you would like him to go to the toilet, if that is outside. If it is an indoor spot (such as a puppy pad or newspaper), he may well find this himself too! He may be sleepy, and not interested in too much investigating initially, and that's fine. Allow time to sit calmly with him.

He will probably be due a meal (see the section on feeding) and should have ready access to water. After eating, and every hour or so, bring him out to the garden again for a toileting opportunity. There is no need to wake him to do so but do bring him out as soon as he wakes up.

Getting used to his new environment is more than enough for him this first day. Dissuade any visitors from dropping by for the first day or two, and just allow him to settle in.

THE FIRST NIGHT

Keep everything calm and quiet before bedtime so your puppy is getting into the right frame of mind to sleep. Remember that stress hormones can stay in the system for many hours, so keeping him calm as much of the time as possible is a really good idea! Let him out for a toilet trip before bed. Don't withhold water – puppies can become easily dehydrated and need access to water at all times. Be sure that the puppy is not overtired or overwhelmed by lots of things happening as it gets closer to bedtime, as that will only make it much harder to settle down and sleep – much like with human children! If your puppy is struggling to settle, a chew before bed can help him get into a calmer frame of mind.

SCOUT'S DIARY – NIGHT 1

'Scout sleeps in my bedroom on the floor (toilet training practicality) in a dog bed. He wakes at 2:30 and 6:30 to go to the toilet.'

WHERE SHOULD PUPPY SLEEP?

Dogs are social sleepers, and as we have mentioned, puppies have a developmental need for sustained physical contact.

For many years, the advice doled out to new puppy parents was to bring their puppy home, put them to bed that night (often in a cage in another part of the house), and then ignore their cries. This, we were told, will avoid 'rewarding' their 'attention-seeking' behaviour and lead to a puppy who can cope with being left alone. The same advice was given to parents of human children for many years, and today many people still adhere to it, for both puppies and human babies.

It has, however, been shown that not responding to infants (human or canine) does the reverse of teaching independence – allowing an infant to repeatedly become distressed in this way is damaging to their ability to establish secure attachments in the long term, and is more likely to lead to clingy, demanding infants, with a deep sense of insecurity which can stay with them for the rest of their lives.

Research has shown us that mammalian brains all work pretty similarly. A lot of the research done on mammalian brains has been done on rats, and we know from this research that there is a period in the 'infant' stage of life where the caregiving that an animal receives has a life-long impact on how prone they are to being anxious. Rats whose mothers were nurturing and caring in their early days (which translates to longer periods in larger mammals like dogs and humans who develop more slowly) had the genes for controlling anxiety 'turned on', whereas those who had 'low-nurturing' mothers never had these genes 'turned on' and suffered from anxiety for the rest of their lives. We have already established that at the age we typically take puppies away from their mothers, they would naturally still be in receipt of a high level of maternal caregiving in the wild. Part of rearing a puppy is providing this caregiving in loco parentis.

We also know that when an animal becomes excessively stressed, the body's response becomes destructive, negatively impacting the brain, emotions, the digestive system, and the immune system. Excessive stress is simply not good for us.

And we know that when a puppy cries, their mum always responds. This all makes perfect sense from an evolutionary perspective. Crying young alert predators to the presence of vulnerable, tasty youngsters! Dogs always do a very good job of rearing puppies, so we could do a lot worse than following their example in this regard.

Mum keeping close and watching over her puppy as he sleeps. Courtesy Jolly Doodles.

SO WHAT SHOULD WE DO?

Dogs find safety in company, and without it they struggle to get the deep sleep they need. All animals that live in family groups will need adults to look after them when they are very young. In these family groups there will always be an adult awake or just dozing, so he can react quickly at the first sign of danger. In a herd of horses, there is always one standing up, ready for action, and the adults will take turns in assuming this role so all of them can get some sleep. Likewise, in a group of dogs or a pack of wolves an adult always remains alert. This provides safety and assurance for the whole group. Puppies cannot take on this responsibility, but when left alone they feel they must, because if they switched off and slept deeply, as they need to, they would not wake up to save themselves from any danger threatening them.

So, the ideal situation is to have your puppy sleep with a member of the family. Most puppies sleep longer and sounder when they are with you, so you might find the night is not as broken as you'd expect.

If you have a carpeted room and are worried about your puppy sneaking off to the corner for a wee (or worse!) during the night, a simple solution is to block off an area of the room for the puppy and cover the carpet in this area with a waterproof bedsheet – these usually have non-slippery cotton on top and a waterproof backing. If you don't want the puppy on your bed, providing access to a slightly raised surface close by can help your dog feel safer and sleep more soundly.

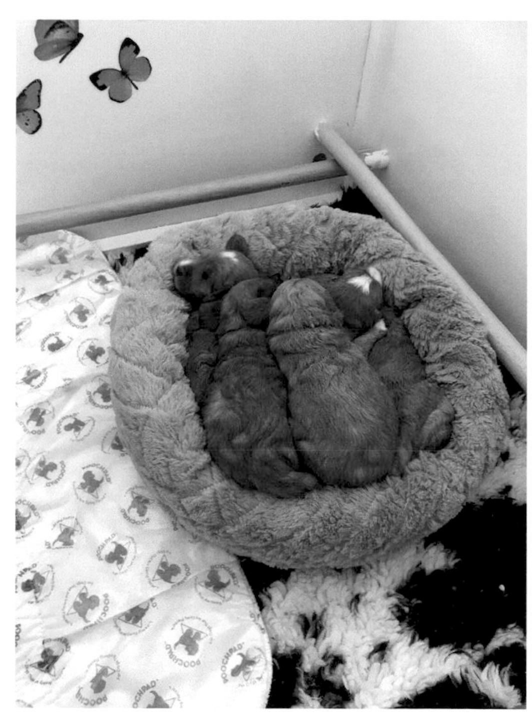

Puppy pile! Your puppy was probably used to sleeping up close and personal with his mum and/or littermates before being taken home. Providing social sleep is really important for your puppy's sense of safety. Courtesy Jolly Doodles.

If you don't want your dog in your room forever, as the puppy gets older you can gradually move their bed further and further from yours. Having their bed just outside your door with a dog gate rather than a shut door can be a good interim arrangement. However, most people seem to quite enjoy having their dog in the room with them, and many choose to allow them access to the bedroom for life.

If they can't be in the bedroom, camping downstairs with them for the first while can be helpful. You can then work on gradually increasing the time they are left downstairs.

SCOUT'S DIARY – NIGHT 2

'Scout sleeps from 10pm to 2:30am to 6:30am, he toilets straight away when let out. I'm hoping I don't get too tired to wake up when he starts moving around and try to avoid accidents in my room. He isn't sleeping at all during the day yet, and is totally overtired, but he must not be able to relax here yet and we have no sleep routine at the moment.'

COMMON ISSUES

The stress of leaving the only home they have ever known, their mother, and their siblings, and coming to a totally new place, will often cause some physical symptoms of stress in your puppy during the first few days. An upset stomach, a minor eye or ear infection, or even a urine infection are not uncommon. These will usually pass within a few days by themselves. Just keep an eye on it, and let the puppy get used to his new home, without creating more stress by taking the puppy to the vet unnecessarily. The more time the puppy has to become familiar with his new home on his own terms, the quicker the symptoms will pass. Of course, if they worsen or your puppy seems very unwell, a vet visit might be required!

THE FIRST FEW WEEKS

It is vital that you avoid overwhelming your puppy with new people or experiences as they settle into their new home. Introduce one new 'thing' a day. This could be a new household item for the puppy to get used to, a new toy, a new human accessory such as a hat, glasses, or walking stick, a new person or a new dog. These new 'things' should be introduced in such a way that the dog can take their time and slowly investigate in a way that's not frightening. If your puppy seems frightened of something, don't force it. Leave whatever it is there and let him decide whether or not to approach. Giving your puppy this choice will help build his confidence and reduce his fear. Forcing the matter will most likely have the opposite effect.

SCOUT'S DIARY – DAY 5

'He is brave about moving around the house, and we want to let him explore at his pace so try to stand back and let him get on with it. We put some items out in the garden for him to navigate around and see the shadows etc. We also leave the bike trailer on the grass for him to see. We are sure to stand well back and not rush him, we want to see how he interacts with the world without our expectations on him (we recently lost a dog whom we did a lot with, like cycle tours, yacht sailing, camping etc, and are being careful not to impose our "recreating" him on little Scout).'

Puppy stands beside toy in garden. Courtesy Georgia O'Shea.

Investigating some boxes. Courtesy Georgia O'Shea.

A treat search over some agility poles lying on the ground can help develop your puppy's proprioception. Courtesy Georgia O'Shea.

Investigating a toy on wheels. This may move if the puppy touches it, and he may be a little startled but as long as he is free to move away or investigate as he chooses, he will be okay! Courtesy Hilde Skatoy.

Checking out a pile of sticks. Courtesy Federica Iacozzilli.

A tentative approach! Courtesy Jolly Doodles.

A polite bum sniff. Puppy checks if this rocking horse smells like a dog! Toy animals often fascinate dogs. Courtesy Jolly Doodles.

Remember that if your puppy has an eventful day, like a trip to the vet for his vaccines, he will need a quiet day or two to recover.

THE VET

First experiences count! Try to ensure that your puppy's first trip to the vet is not for scary ouchy vaccines.

Many vets are very facilitative about people bringing their puppies in for a meet and greet, to be given treats by the vet nurses and to head off again thinking the vet is a wonderful place!

However, think twice about puppy parties in vets (or elsewhere!) where large groups of puppies are brought in to play madly for an hour. These situations often result in some puppies learning to bully, and others being bullied and learning that other dogs and the vets are both very scary!

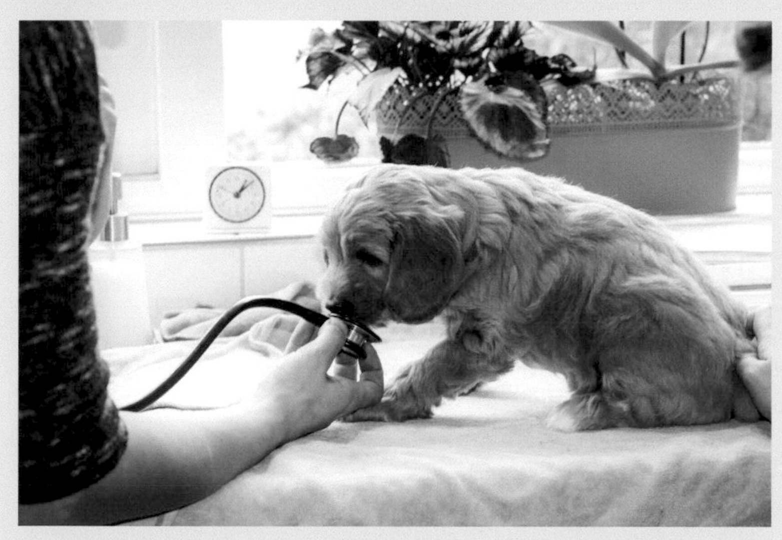

Making sure your puppy's first experience at the vet isn't overwhelming or frightening is important. Courtesy Jolly Doodles.

PAWS FOR THOUGHT: *SAUMFARING*

Saumfaring is a Norwegian word for a thorough checking of something (literally seam checking!), and we just like it, so we are sharing it with you!

It involves a quick physical check of your dog, looking for anything out of the ordinary, following a certain pattern.

Place one of your palms on each side of your puppy's head, and gently move them slowly, with no pressure, from muzzle to tail, along the face, the neck, the back. Then move them along the sides of the puppy, the chest and stomach areas, and down each of the four legs.

The whole procedure takes maybe 20 to 30 seconds, and you can easily feel any changes or differences as your hands move. Lumps or bumps? Areas of heat? Colder areas? Tension in the muscles, an unusual reaction to touch, or anything else strange. Muscle tension might mean a muscle has been overworked, which can easily happen. A sore muscle is a clear sign that there has been too much of something. Hot spots are often the same. Maybe the puppy needs a little rest, a massage, less exercise, or a different type of exercise. Regardless, signs of muscle pain indicate that something needs to be looked at.

Turid has done this daily for nearly 50 years and knows for a fact that identifying the need for help at an early stage saved the lives of a couple of horses and dogs as well.

So make it part of your daily routine, and start when your puppy is young. Most people touch and stroke their dogs often anyway, but doing it with a purpose is different, and makes you feel the dog in a different and more conscious way.

TAKE-HOME POINTS FOR CHAPTER 2

- If possible, take your puppy home at 10 to 12 weeks; try to avoid the eight-week fear period.
- Your puppy is wired to still require high levels of maternal care-giving. Don't expect too much independence at this early stage.
- Keep things quiet and calm and slowly introduce new things, people, and activities. Initially, his only activities should be having the freedom to explore the house and the garden and the local environment. Just let him explore and get used to his new environment at his own speed.

3 *Meeting your puppy's needs*

SAFETY

Feeling safe is crucial to any living creature's well-being. For a puppy, feelings of safety come primarily from being close to someone who will care for them and protect them from dangerous situations. It is crucial to protect your puppy from situations where he may feel unsafe, and let him know that he can rely on you to help him out if he's feeling frightened or overwhelmed.

An absence of pain or punishment and the threat of pain or punishment is another critical element to your puppy feeling safe (see more on punishment on page 78).

Additionally, many puppies take great comfort from routine, and like to be able to predict what's going to happen next. If you feel your puppy is spending a lot of time looking at you, this is probably why! Keeping things predictable while your puppy is settling in can help build a sense of safety and security.

Another thing to consider in terms of creating a feeling of safety is respecting your puppy's boundaries and space – for example, not disturbing him while sleeping or eating, not allowing children (or adults!) to crawl into his bed, and not grabbing things from him.

SLEEP

Puppies need sleep, and lots of it! A puppy needs as many as 20 hours of sleep a day, so you won't be surprised to hear that many of the puppies we meet are sleep-deprived!

Dogs are also polyphasic sleepers, so it is normal for them to get this sleep in many blocks in a 24-hour period. An adult dog will get about 40% of his sleep during the day.

As we have already mentioned, dogs are also social sleepers. In order to get the deep sleep they need for growing and memory consolidation, it is important for dogs to have access

Adult dogs sleep for 12 to 16+ hours a day. This means that in the wild, the puppy would have more time when the adult dogs were resting, and would probably follow suit. As we sleep about 50% less than dogs, and puppies like to keep an eye on what we're doing, they often don't get as much rest as they need.

It is believed that dogs need these additional hours of sleep because they spend less of their sleep time in REM sleep. Humans spend about 25% of their sleep in REM sleep, whereas dogs spend just 10% of their sleep time in REM sleep.

DOI: 10.1201/9781003305156-3

Social sleeping can be achieved with another dog or with a human. Most dogs will choose to sleep near their humans if given the choice. Courtesy Georgia O'Shea.

to their social group. Throughout your puppy's life, allowing them access to their canine or human companions while they sleep is a good idea. However, this is particularly so for puppies.

When you go to sleep, allowing your puppy to sleep near you can also bring a number of benefits to the human part of the equation. People who try to isolate their puppies will often suffer from many broken nights' sleep while the puppy howls mournfully for them, or barks at every little noise. Once the puppies are brought into the bedroom, or the human joins them downstairs, they will generally sleep soundly.

Being near your puppy at night can also help speed up toilet training, as the puppy can alert you when they need access to the toilet.

Lastly, dogs are crepuscular in nature, which means they are naturally most active at dawn and dusk (the best times for scavenging in the wild!), so if you find your puppy is up and ready to go first thing in the morning, and tearing around the house in the evening, that's why! The good news is that most dogs are pretty good at adjusting to our circadian rhythms, and in time, most puppies will sleep a bit later.

PRACTICAL SLEEP SOLUTIONS

For dogs, the ideal sleeping place is often up high, as this provides them with a sense of safety and security. They like to be close to others, and need to be able to stretch out. Think about how you can best accommodate your puppy's sleep requirements given your own situation.

With you in/on the bed: This is the preferred option for most puppies. Accidents are rare once the puppy realises that the bed is his sleeping area too! A lot of puppies will never pee on the bed. You can help set them up for success by bringing them to your bed for the first time soon after they've been to the toilet, and then scattering a few bits of food on the bed (see toilet training tips). But if you're worried, invest in a waterproof mattress protector for the first few weeks and don't use your best duvet!

Contrary to the old wisdom, allowing your dog on the bed or sofa will not give them dominance issues. Dogs wish to be on sofas and beds because they're high, cosy, and smell of us. It has nothing to do with a desire for world domination or to subjugate you!

Scout stretches out on Georgia's bed. Height and comfort combined with his human's smell makes this a top spot for sleeping soundly. Courtesy Georgia O'Shea.

On a raised bed beside your bed: Raised dog beds can be purchased, or you can improvise with half a cardboard box with a cushion inside on an armchair beside the bed (just make sure your puppy can't jump out and cause himself an injury). If your puppy gets upset during the night, you can reach out and provide some physical contact.

On a dog bed in your room: If this is all too close for comfort, sectioning off a part of the bedroom where they can freely move, putting a waterproof cover down within the sectioned off area as well as the puppy's bed, and a toileting area (some newspapers or a puppy pad) may be a workable compromise! The dog will still be able to hear, smell, and see you, but you will have some distance from him. This space should be big enough to allow the puppy some freedom of movement, and to have a distinct space for going to the toilet should he need to.

Scout was promoted to bed sleeping once he was doing well on the toilet training front. Courtesy Georgia O'Shea.

A raised dog bed. Courtesy Georgia O'Shea.

CHOOSING BEDS FOR YOUR PUPPY

You will most likely need a number of beds for your puppy to facilitate both night-time sleep and daytime naps! Bear in mind that dogs need to be able to stretch out to get their REM sleep, so your dog should have access to beds big enough to facilitate this.

There is a really wide range of dog beds available, and dogs will have different preferences for different types of bed. Some will like doughnut-style beds where they can rest their heads on the raised edges, others will like 'igloo' beds that they can hide away in. Dogs often like to sleep up high, so your puppy may prefer a raised bed (or the sofa!), and the less hairy breeds of dog, such as greyhounds and whippets, may appreciate 'snuggle beds' with a blanket attached that they can get under for extra warmth!

However, a thick duvet folded in four is always a really popular dog bed hack!

ENCOURAGING SLEEP DURING THE DAY

Your puppy will need periods of sleep during the day as well as during the night. You'll probably find that your puppy wants to be involved with everything you're doing and simply isn't racking up the necessary sleep hours during the day. If you can, set aside periods of time where you can sit still, read a book, or do some meditation! You can try doing some basic, calm, scent work (see page 56) with your puppy or give him something to chew before these naps to help him relax.

COMMON SLEEP ISSUES

My puppy isn't sleeping enough but I just can't get him to settle down. He is always full of energy.

Three Cs: company, comfort, calmness

If you're having problems getting your puppy to get enough sleep, think of the three Cs – company, comfort, and calmness.

Company

As we've already discussed, puppies cannot sleep well when they are left alone. So, the first thing to do is to lie beside him, let him sleep beside you, or on your lap, in your bed, whatever feels best. And sit still, take the time to be with him, and he will sleep.

When they get a little older, they usually feel safe enough in their home to be able to sleep with more distance, but preferably still being able to see and hear that someone is in the house, taking responsibility. As adults, they often learn to feel safe enough to sleep alone, but as puppies, they cannot.

Comfort

Much like humans, a puppy's environment will affect his ability to sleep soundly. Puppies can get cold easily, and then they can't sleep. Likewise, if they are hungry or thirsty, they will not sleep soundly, and will wake up often. They also need to feel comfortable in their bed, so let your puppy choose where he wants to sleep, and provide blankets and cushions for him to snuggle up in should he wish.

Calmness

The next area to look at is their routine. Vet behaviourist Amber Batson has wisely said 'calmness is a way of life, not a learned behaviour', and this is a model we subscribe to as well! If you are spending lots of time playing exciting games and winding up your puppy, being excited and wound-up will become his default way of being. However, if your puppy spends most of his time engaged in calm, more thoughtful behaviour, this will become his default way of being – and that will make your life an awful lot easier! Everybody who has had children in the house knows that if we play exciting games with them close to bedtime, they get over-excited and can't sleep. Or, if there has been too much happening during the day and they are overtired, they might also have problems sleeping. Puppies are no different. The activities we do with puppies should be short and in moderation – with as little excitement as possible and in short sessions only.

It is important to avoid lots of rough play, throwing balls, sticks and frisbees, taking them to huge groups of dogs in the park, puppy classes that last too long and demand too much of them. These are traps that we can easily fall into with a puppy, and which can negatively impact on their sleep.

We have to look at all these different aspects of their routine to find out why the puppy is not sleeping, and seems to be overactive. If you feel he is overactive, this is a clear indicator that he is stressed. There are always reasons for stress, which can usually be found in the absence of one of the three Cs!

MOVEMENT

Until a dog's growth plates have fully closed, we need to keep a close eye on how much repetitive exercise (including going for walks or runs) our puppies are doing. This is part of the reason we recommend short walks for puppies. The rule of thumb is five to ten minutes of on-lead walking once or twice a day at three months, and then add about five minutes on for every additional month.

Walking should be slow, and sniffing encouraged! Slow, deliberate movements will develop your puppy's deep muscles and provide his skeleton with the stability it needs.

Of course, your puppy needs more general movement too, so here is some guidance for ensuring your puppy gets everything he needs in terms of exercise for those first few months:

1. When walking on the lead, make sure you're walking slowly enough that the puppy is actually walking and does not need to trot or run to keep up. He should be free to stop, sniff, and run a few steps if he chooses. Do not ask him to look up at you as this will put his neck and spine under stress, and do allow him to be on either side of you – always keeping him on one side will create an imbalance in his body. A variety of movement is necessary for your puppy's development, so his walks should be more than just repetitive walking in a straight line. Having a long-enough lead to facilitate this movement is crucial.
2. Find safe places where the puppy can run free for 20 or 30 minutes (although perhaps not a dog park! Read more about some of the pitfalls of dog parks on page 42). Because the puppy will move at different speeds, can stop to sniff whenever he likes, and can take breaks as he needs to, this sort of exercise is different to the sort of repetitive movement your puppy would be doing walking or running on-lead. Young puppies do not tend to run away from you, but if you are nervous, you can choose

Tirian the puppy stops on his walk to observe the birds. Allowing puppies time to take in their environment is really important. Courtesy Therese Asp.

Having the opportunity to explore is so important for puppies. Courtesy Therese Asp.

a fully fenced area. It is great if the area has some variety of terrain – little hills and different surfaces – as this will help the puppy develop a variety of muscles. When your puppy is enjoying this time off-lead, just let him move around exploring without interfering or asking for specific behaviours. This will allow him to rest when he needs to and to move freely, and will help you learn to trust your puppy!

3. Allow the puppy to move freely around the house and garden – all the time! Locking a puppy in a crate deprives him of the opportunity for this regular movement which is so important for his development and well-being.
4. Create little physical challenges for the puppy – like a low ramp, sticks on the ground that he can walk over, a short tunnel. There should be no commands or pressure – just watch how he gets curious and gives things a go, and how quickly his co-ordination and proprioception (an awareness of where one's body is in the world and how it moves) improves. Spending a few minutes on these sorts of exercises daily will pay dividends!

More exploring! Courtesy Darja Fir.

Setting your puppy some gentle physical challenges that he can attempt in his own time and at his own pace can help with his physical development. Courtesy Hilde Skatoy.

FETCH

We implore you to *avoid* playing fetch with any dog, least of all a puppy! There are many issues with this game, including:

- The physical aspect – sudden accelerations, stops, twists and turns, and landing on the hind legs after a catch, all put your puppy's body under enormous strain, and risk doing serious long-term damage (which will incur expensive vet bills!).
- Dogs can become obsessive about the game and become addicted to the dopamine that's released when playing chasing games, and react with signs of stress when ball play is interrupted or ceases.
- When your dog bounds after that ball, it's the prey/chase response kicking in. And with that comes the fight or flight hormones, like adrenalin, and later, cortisol. Adrenalin rises and falls relatively quickly, but cortisol rises and drops more slowly. It can take hours for your dog's stress hormone levels to return to normal levels, and if another stressful incident takes place in the meantime, those hormones, particularly cortisol, will begin rising from an already elevated level, and take even longer to reduce.
- The dog can end up in a state of chronic stress, and recovery from this state (once the stressors have been removed) can take close to a year. A chronically stressed dog can suffer from a compromised immune system, reactivity, poor social skills, skin disorders, stomach issues, and many other problems.

Find some fun alternatives in our mental stimulation section!

TROUBLESHOOTING

My puppy has mad bursts of energy

Many puppies will have periods of excitement during the day (often dawn or dusk) when they want to run around and burn off some energy. This is fine. It's better if this activity can take place on surfaces (such as carpets or grass) where they are less likely to slip and slide. Allowing them to do this without too much human input will help ensure that the movement is natural, and not too repetitive. Remember that any sort of chasing is going to be repetitive, high-impact, and very hard on their joints.

My puppy doesn't want to go for walks

People are often concerned that their puppies don't want to walk. This is quite common in puppies, and the usual reason for this is that it's simply too much for them. For a period of time, try driving or carrying him to a place where he can just run free and choose what do to.

In some cases, owners are walking too fast and the puppy has to run constantly to keep up. As you can imagine, this is not a pleasant experience for the puppy, and as a result, they don't have positive associations with going for walks.

As they hit puberty, their confidence grows, they get stronger and their sense of smell becomes more developed, and the outside world becomes more interesting. In the meantime, just go with it. Don't try and bring them too far. You can also help make the world more interesting by scattering some food on their walks for them to find, pointing out things that will smell interesting, or walking with a nice dog friend.

The world can also be a scary place for a young dog, so if your puppy seems afraid, don't force the walk. If he just wants to go as far as the front gate initially, that's fine. Let him investigate and experience things at a level he's comfortable with and in his own time, and his confidence will grow.

My puppy is very restless after his walk and won't settle

This may be counter-intuitive, but a restless puppy most likely means you've overdone the walking, and not the reverse. Puppies can be a bit like toddlers, and becoming too tired or overstimulated can lead to them being in difficult humour and difficult to settle. Just allow your puppy to take it easy for the rest of the day, and have a quieter day tomorrow!

Puppy Tirian takes a break to rest and watch. Courtesy Therese Asp.

My puppy bites the lead on walks
This is a clear sign from your puppy that it's all become too much and that they're overwrought! It was time to head home five minutes ago. If it's at the beginning of the walk, take a look at what's been happening earlier in the day and the previous day. Has the puppy had enough sleep and rest time? Has someone been winding them up? Has something stressful happened? Are they nervous or overwhelmed on their walk? Either way, it's time to go home and rest and take things a lot easier the next day.

My puppy picks things up on walks
Puppies investigate the world with their mouths, and it is normal for them to pick up things they find when out and about. A mistake we commonly make is to grab every single thing they pick up out of their mouths. Paying so much attention to this activity makes them more likely to keep doing it, and more likely to swallow things if they fear we are going to 'steal' them. If the item is unlikely to cause your puppy harm, try to do nothing. Usually, they carry these things for a few seconds and then spit them out again. Of course, if the item is dangerous, you may need to intervene (preferably by swapping it for a toy or treat rather than grabbing it!), but if it is something harmless, leave the puppy be and wait for them to grow bored of it and spit it out – it usually doesn't take long!

FOOD

Choosing food for a dog can be a bit of a minefield. There are new brands and different types of dog food popping up all the time. As we have already mentioned, our suggestion is that you stick with whatever your breeder was feeding the first few weeks and use this time to research the sort of food you would like to feed your dog.

HOW OFTEN?

Puppies need to eat more frequently than adult dogs, and we would recommend that a puppy eats at least four to six times a day until he's about four months of age (unless they decide themselves to eat fewer meals). They should then eat at least three meals a day until they're around six months, and then at least two meals a day for life. Some people find their adult dog does best on three meals a day rather than two – this can be particularly true

FOOD FRUSTRATION

Dogs are gobblers by nature. It's normal for them to eat quickly and trying to prevent them from doing so by using slow feeders or hand-feeding throughout the day is likely to cause frustration and possibly long-term issues with food.

Main meals should contain a sufficient quantity of food, and the food should be easily accessible (e.g. in a bowl). Enrichment activities involving food should be in addition to meals, and not done on an empty stomach.

for highly strung or reactive dogs, as it can help keep their blood sugar more steady. If you are feeding three meals, just divide their daily allowance by three instead of two, so that you are not over-feeding.

HOW MUCH?

The amount of food required by your puppy will depend on their size and the food they are eating. Food with lower nutritional value will often necessitate feeding larger portions for the dog to get the nutrients they need. Higher-quality food will usually require smaller portions. Feeding guidelines are provided by the food manufacturer, and this is a good place to start. However, it is not uncommon to meet underweight puppies. Puppies are a bit like humans in that their body shape changes as they grow into adults. Puppies should not necessarily be the same shape as an adult dog, and a bit of puppy fat never worries us. It is also not uncommon for a puppy to be a very healthy weight, and then suddenly take a growth spurt and need to eat more to maintain that healthy weight. This can be particularly hard to notice in fluffier breeds, so feel your puppy's frame regularly. As a rule of thumb, you should be able to feel the ribs but your fingers should not be able to fit between the ribs, and you should not be able to see vertebrae or pelvic bones.

HOW?

Eating should be a stress-free experience for your dog, and as a basic survival need, food should be provided unconditionally. We don't recommend asking your dog to sit or wait before feeding, or making him 'work' throughout the day to earn his food as rewards. Knowing that his basic needs will always be met, no matter what, will help create a happy, confident dog.

Did you know that dogs are social eaters and prefer to eat with their social group around?

Try sitting down in the same room as your puppy while they're eating, and help make dinner time even better for your dog!

Social eaters

Never try to take your puppy's food away from him. While it used to be said that this would prevent resource guarding (when a dog displays behaviours such as growling, snapping, or biting if they think anyone is approaching a valued resource such as a toy or food), it is now widely acknowledged that is more likely to have the reverse effect! It simply teaches the dog that it is not safe to have humans around food.

Creating a sense of security and predictability around food is one of the best things you can do for your puppy – practice abundance, and they are unlikely to have food anxiety!

VARIETY

It is normal for dogs to eat a wide variety of food. Free-ranging dogs are predominantly scavengers, and on a normal day will eat many different types of meat and vegetable matter, as well as waste products.

Pet dogs, on the other hand, are often expected to eat the very same kibble, twice a day, for the whole of their lives... If they tire of it, they are often labelled 'picky eaters'.

> **A VARIED DIET**
>
> One study of free-ranging dogs in Africa found that their diet consisted of roughly 53% various animal/insect protein, 24% various vegetable matter, and 21% human and animal faeces...

However, you can use food variety as a great way to enrich your dog's life. Give them the opportunity to taste new things, add fresh meat to their meals, and offer them vegetables to taste and chew. Adding fresh food to their diet is also a great way to boost their nutritional intake, just be sure to avoid anything toxic to dogs!

Table 3.1 Showing Some Foods That Can Be Added to a Dogs' Diet and Food That Should Be Avoided. This Is Not an Exhaustive List – Do Check Before Feeding Your Dog Anything New That It Won't Do Them Harm

Some healthy foods to add	Foods that are toxic or should be avoided
Carrots	Chocolate (especially dark)
Celery	Xylitol (sweetener, often present in sugar-free items such as chewing gum)
Cucumber	Raisins and grapes
Meat (cooked or raw – however, if feeding raw pork, some advise freezing it in advance to kill certain parasites)	Macadamia nuts
Cooked/previously frozen fish	Alcohol
Eggs	Cooked bones (can splinter)

FEEDING OPTIONS

There are many different feeding options available and a wide variety of opinions on all of them! We are not nutritional experts, and suggest you consult a vet with a special interest in nutrition to get the best advice on feeding your dog. However, in this section, we will share our own experience of feeding dogs, and direct you to some useful resources.

No matter what you decide to feed, we suggest you buy as high quality a food as you can afford, and supplement with tasty fresh food for variety and a nutritional boost!

KIBBLE

Kibble is a convenient food for many dog owners to feed their dogs, as you can buy in bulk, don't need fridge space for it, and it is usually the most affordable option. When choosing a processed dog food such as kibble or tinned food, it is worth asking yourself the following questions:

- Do I know what every ingredient listed actually is?
- Is the primary ingredient identifiable meat/meats? It is usually a good sign if the first ingredient is meat, poultry, or fish of some sort. 'Meat meal', or 'animal derivatives' could include hair, teeth, feathers, and other animal by-products of minimal nutritional value. As with human food, the manufacturers have to list the biggest component ingredient first. But beware of clever marketing that lists a meat ingredient first, but then numerous types of starch individually, e.g. 'poultry protein, maize, rice, wheat, maize gluten…' A percentage can be a useful indicator of how much meat is actually present in the food, e.g. 'fresh turkey (60%), potatoes, peas…'
- Does it contain unnecessary additives, colourants, or preservatives? We've all heard about the concerns that certain E-numbers can have on children's behaviours, and there is research that indicates they can have a similar effect on dogs' behaviour. Flavour enhancers can also encourage dogs to eat their own faeces – a behaviour most of us would rather our dogs refrained from.

WET FOOD

Wet food is also easily stored, usually coming in tins or pouches, and is usually more expensive than kibble. Many dogs will prefer wet food to dry kibble. As with kibble, there is a wide variety of wet food available, of differing quality and at different price points, and the same considerations should be applied when choosing a wet food as when choosing a kibble.

REGURGITATION

In the wild, adult dogs regurgitate food for the puppies up until they're four to five months old. As you can imagine, regurgitated food is much softer than the hard kibble we often feed our puppies, which can be painful for them to eat.

RAW FOOD

Raw food has become very popular and is seen by many as a less processed and more natural way of feeding dogs. Many people buy commercial raw food, which is a complete food, usually ground up and frozen. Cost-wise, it tends to come out at roughly the same price as a premium kibble. Other people make their own, using off-cuts from the butchers or meat bought in the supermarket, which is usually a cheaper option. www.dogsfirst.ie, a website run by canine nutritionist Dr Conor Brady, has really useful resources for anyone considering going down the raw route, including a feeding guide. There are a number of breeders who wean their puppies directly onto raw food, so it is not necessary to wait until your puppy has grown up to change to this option. Not everyone has the budget or the stomach for raw, however, and it does require fridge and/or freezer space for storage.

HANDY FOOD RESOURCES

www.allaboutdogfood.co.uk – An independent dog-food comparison website
www.dogsfirst.ie – Raw feeding website run by canine nutritionist Dr Conor Brady
Feeding Dogs by Conor Brady
Raw and Natural Feeding by Lew Olsen

FRESH, COOKED FOOD

A relative newcomer, fresh cooked food delivered to your door is growing in popularity. For many, this brings all of the benefits of a fresh, minimally processed diet for their dog, without the 'ick' factor of a raw diet. Again, fridge/freezer space is required, and this is often the most expensive option for feeding your dog. Other people prepare their own dog food (which can be more cost-effective), and it is possible to find recipe books containing balanced meals for dogs.

A 2003 study showed that dogs fed 'real food' live on average 30% longer than those fed commercial dog food (Lippert and Sapy 2003).

Batch-cooking fresh food for a puppy. Courtesy Georgia O'Shea.

Please do bear in mind that just because a dog food is sold at your vet's, this doesn't mean it's good quality. In fact, it often means that you're simply paying above the odds for a very

SCOUT'S DIARY – DAY 7

'We introduce him to Yakers chews to help his teething. He has tried some puzzle toys we had been gifted but we use these for extras when he has already had his meals (chicken, cheese etc.) as we want him to be content with food and not encourage begging or being obsessed with food.'

Scout enjoying a Yaker. Courtesy Georgia O'Shea.

average food. Always look at the ingredients and make an informed decision. Some vets are very knowledgeable about canine nutrition and will have informed themselves about the various options out there. Others will have only received a few days' training in canine nutrition during their time in vet school, and that training may have been sponsored by the big pet food manufacturers.

COMMON QUESTIONS

Should I leave the bowl down?

People often ask whether they should pick the puppy's bowl up if he doesn't immediately eat the entire contents of the bowl. If the food is fresh or a wet food, and likely to go off, it is advisable to pick it up after an hour or so at room temperature. Kibble can be left out for longer. There is no harm in your puppy not feeling hungry and deciding to come back for their food later. If it is not possible to leave the food down for hygiene reasons, you can always put it in the fridge, and if it looks like your puppy is looking for it, put it out again.

If your puppy is routinely leaving food, it is worth asking why. Are you giving him too much? Does he like it? Is he bored of it? Is it difficult for him to eat?

My puppy is getting picky

'Picky' is often a word used to describe a dog who has tired of eating the same food, numerous times a day for weeks on end. This is perfectly normal, and the same would happen to any of us. As we have said, it is normal for dogs to eat a really wide variety of food. Try adding different flavours and textures to the food, or varying the type of food. If your puppy is inappetent and refusing to eat anything at all, it may be time to see the vet.

My puppy is constantly hungry

Dogs, like humans, have varying appetites. With very hungry puppies, check their weight and make sure they're not too thin, and that you are feeding them frequently enough. You may also want to rule out parasites.

If your puppy is in good health, getting adequate quantities of food but still looking for more, you can try offering healthy, low-fat snacks such as apples, carrots, cucumber, celery, etc. Giving your puppy a food-based chew after their meals can also help them feel satiated.

My dog had loose stools – is it his food?

Severe diarrhoea can become a medical emergency in a puppy, as diarrhoea can quickly cause dehydration. It can also indicate a parasitic infection such as worms or giardia. However, the occasional loose stool is as likely to be linked to stress or excitement as diet or illness. Before blaming the food, take a look back over the previous day or so, and see if anything out of the ordinary happened that your puppy might have found scary, stressful, or exciting.

Sudden changes of food can also cause an upset stomach, so make sure to introduce a new diet or type of food slowly, over a period of about two weeks to minimise the risk of this.

Can I give my puppy leftovers?

So long as they are not dangerous to dogs, it's fine to give leftovers to your dog. If you don't want to encourage begging, you can give them the leftovers away from the table.

My puppy is eating his poo (coprophagia)

This is another reasonably common complaint. There are a number of reasons why this can happen. The first thing to rule out is hunger! Ensure your puppy is eating enough and that the food is of good nutritional value. Physical problems may also contribute to this behaviour. One theory that has been put forward is that if the puppy has digestive issues, and the food has not been properly digested, it can smell very similar coming out to how it did going in! As we've mentioned, feeding a food with a lot of artificial flavourings may also lead to faeces eating. From a behavioural perspective, it has been theorised that puppies will have witnessed their mother eating the puppies' faeces to keep the den clean, and that they are simply copying this behaviour. This also happens quite frequently with puppies who grow up isolated from their mother, in small places like a crate. Puppies like their living area to be clean, so if mum is not there to dispose of the mess, the puppy might start doing it himself, and carry on doing this even when he no longer lives in a crate.

Another theory is that faeces-eating can occur when a puppy has been told off for going to the toilet indoors. They can learn to associate the presence of the faeces with human anger and so try to dispose of it.

Rule out any physical issues with your vet, and make sure to promptly remove any faeces so this does not become a habit. Some people swear by adding a teaspoon of pineapple juice to their dog's food, which should not affect the palatability but apparently doesn't smell very appetising once it's been digested and eliminated.

CASE STUDY: BARKING BRAN

When Bran was about five months old, his humans told Steph that he was barking at the family while they were eating their dinner. Steph suggested they feed him before they ate so they could be sure he wasn't hungry at this time. They got back to her, and said that even though they were offering him his food, as well as something to chew, he was eating neither, choosing rather to keep barking at them. It appeared he was not barking because he was hungry... However, things are not always as they appear!

His humans then tried feeding him some cooked chicken and rice before eating themselves. He gobbled this up, and the barking stopped. So it seems that Bran was hungry, perhaps struggling to eat his hard kibble, and that finding something he enjoyed and ensuring he was satiated was all it took to change this behaviour!

Bran checks out what the humans are eating! Courtesy Shane Ó Cathasaigh.

PAWS FOR THOUGHT – FOOD

Do some research into the sort of food you're going to feed your puppy. Check ingredients, cost out the various options.

Also, think about how you're going to manage your delivery of the food so that your dog can eat in a way that is compatible with his natural behaviours:

- Eating substantial meals in peace.
- Providing variety.
- Ensuring there are things to chew.

Constant access to water is a basic puppy requirement. Courtesy Jolly Doodles.

WATER

Dogs, and especially puppies, need access to water at all times. Do not be tempted to take away your puppy's water to try and manage their toileting patterns. If your dog seems to drink frequently and you are struggling to figure out when they are likely to need to go to the toilet, you can try switching to a food with a greater moisture content. If they are getting more moisture in their food, they can be less likely to graze on water throughout the day. It is also worth checking the salt content in the food. Water should still be available at all times.

ELIMINATION

Elimination is a basic need for any animal. Yet, it is often observed that pet dogs are the only animal that is not allowed to choose when to go to the toilet. Imagine how stressful it must be for our dogs to need to go to the toilet and not to be able to. And worse still, not to know when someone is going to come and let you go to the toilet.

Before male puppies begin to lift their leg to urinate, they will lean forward like this. Courtesy Jolly Doodles.

Providing your puppy with free access to their toileting area is best, and in the early days, can really help speed up the toilet training process. In the summer, you may be able to leave doors open. A dog flap may also be an option. Where this is not possible, watch for signs your puppy needs to go, and allow ample toilet breaks! If they don't want to go, don't push it, just try again a bit later.

The main focus of your toilet training should be on giving your puppy plenty of opportunities to get it right, and observing his behaviour for hints that he needs to go (see page 91). We do not recommend putting your dog's elimination on cue. As we have said, they will want to be clean as soon as they can, and will try their best to be clean. Our side of the bargain is to offer them ample opportunity to eliminate so they are never forced to hold it for too long).

SOMETHING TO THINK ABOUT...

It is easy to feel irritated when you come home and find an accident on your best rug.

However, before you get cross, try and think about it from your dog's perspective. Perhaps you've been gone for four hours, and your dog realised he needed to go to the toilet an hour after you went. He held on for another two hours, feeling very uncomfortable. He had no idea when anyone would come to let him out. He was probably in quite a significant amount of pain and stress by the time he gave up and went in the house.

However, despite being put in this difficult situation, your dog will in all likelihood hold no grudges and be delighted to see you.

Is it really reasonable for you to be cross with him?

MENTAL STIMULATION

Providing your puppy with mental stimulation is necessary for their development and well-being, and is a great way of tiring them out and buying you a bit of quiet time! Mental stimulation is anything that gets the brain working, so can be situations that get them thinking, challenges to overcome, or new environments to get to grips with. However, one of the easiest ways to engage their brains in a way that's fun for both you and them is by giving them opportunities to use their noses.

As sight-centred humans, it's easy to forget how nose-centred our dogs' existences are.

The human nose is actually much better than most people think it is, and humans can distinguish thousands of different smells even in minute quantities. But dogs' noses are truly incredible, and our sense of smell pales to insignificance in comparison. To put the difference in context, we have five to six million odour receptors, while dogs have 220 million. The olfactory epithelium of dogs is about 20 times the size of that of humans. It is perhaps no surprise then, that dogs can detect substances at concentrations of up to a million times lower than humans can perceive them.

There are so many reasons to give your dog plenty of outlets for using their noses, including:

- Nose work is a great way to build concentration, focus, and self-confidence in your dog.
- Doing 10 to 20 minutes of intense scent work leaves a puppy pleasantly tired.
- Sniffing causes a dog's pulse rate to drop, making it a nice calming activity.

Want to find out more about scent games? Anne-Lill Kvam has a brilliant book about how to teach your dog all sorts of great nose work games, called *The Canine Kingdom of Scent*.

SCOUT'S DIARY – DAY 32

'Today Scout tried a licki mat for the first time. He tried to employ the same tactics on it as he does the other food toys (throw them around) and then quickly changes tactic. It is fascinating to watch him think through the process of how to resolve it. We are keeping all the challenges he faces really achievable so he is confident about problem-solving.'

Scout enjoying a licki mat. Courtesy Georgia O'Shea.

- Sniffing provides great mental stimulation for dogs, something dogs we meet are often lacking.
- Playing nose games with your dog is a great way to positively interact with your dog, and to increase your value on walks – who wants to go too far from the human who may at any moment toss a handful of treats into the grass for them to find or engage them in some other fun scent task!

SOME WAYS TO PROVIDE OLFACTORY ENRICHMENT

- First and foremost, allow your puppy to sniff on walks. Your dog's walk is probably one of the highlights of their day. Frog-marching them around the block, pulling them along every time they try to stop and sniff is akin to bringing you to an art gallery and covering your eyes every time you tried to look at something! The sniffing will tire them out as much as the walking will, so worry less about the space covered, and more about the quality of the walk.
- Do a simple treat search – scatter lots of little pieces of food around for the puppy to find. You can use the grass in your garden, piles of leaves in the park, a thick-piled rug, or a specially made snuffle matt. Remember not to help – it's the searching that the dog finds most rewarding, not the finding.
- Put treats/pâté/fish paste on various objects on walks, and leave your puppy to investigate – try trees, fallen logs, and along low walls.

Searching for pâté on a tree is a good way of encouraging sniffing.

Searching for treats on a fallen log improves balance while also providing mental stimulation. Courtesy Georgia O'Shea.

- Create an enriched environment for your puppy to investigate (see page 124).
- Hide treats in blankets, towels, and laundry.
- Turn 'fetch' into a much calmer game of 'find and retrieve' by hiding the toy instead of throwing it.
- Drag a sausage along the ground, and hide it at the end of the trail.
- Teach your dog to use his nose to find your keys, his toys, and even hidden people!

Remember to do these food-based activities relatively soon after your puppy has eaten. A hungry puppy can find this sort of activity incredibly frustrating, and the release of gastric juices in an empty stomach which will not imminently be in receipt of a significant amount of food can cause discomfort.

Social interactions are also mentally tiring for puppies. Introducing short, calm, on-lead walks with other dogs will allow your puppy to practice his social skills, whilst also providing plenty of mental stimulation.

SCOUT'S DIARY – DAY 13

'Today Scout meets my partner Hugh's two salukis, who are both very neutral calm dogs, and explores their garden. Hugh brings him home after a short while as we don't want him to become overtired or worried, although he seems to be fine. Scout is exhausted afterwards; it reminds us how much of his energy it takes just to cope with a new experience.'

Scout meeting the in-laws. Courtesy Georgia O'Shea.

PAWS FOR THOUGHT – MENTAL STIMULATION

Commit to making mental stimulation a part of your dog's daily life. Plan ahead, and decide what you'll do each day for a week. Remember, this can be a slow sniffy walk, a simple treat search in the garden, an enriched environment, or something more complicated!

FOOD FOR THOUGHT...

A word commonly used in dog training is 'motivation'. It is often said that a dog has to be motivated to work with you/for you. If they won't do what you want, they're not motivated enough. Motivate them more! Use better treats. Better treats not working? Take away their meals and hand-feed them throughout the day in return for them doing what you want.

Now, call us naysayers if you will, but we believe that this idea of 'deprive and reward' is not a kind way to interact with your dog. And it's not 'positive' dog training.

EATING = SURVIVAL

The reason dogs are motivated by food is because they need it to survive. Anything we need in order to survive is going to be inherently valuable to us. To only fulfil that basic need conditional to their compliance with our every wish is not fair.

Having to earn every mouthful of food means the dog has to be constantly on alert, just to ensure they get this basic need met.

WHAT ARE YOU ASKING YOUR DOG TO DO AND WHY?

If the only way you can convince your dog to do something is to compel them to do it (essentially by threat of starvation), why is that? Some of the most common reasons for training problems are mentioned below, and in any of these situations, just increasing the value of your rewards until the dog complies (and he probably will eventually comply – we all have a price, especially if we're constantly hungry!) is not going to benefit the overall well-being of the dog, and could in fact, adversely impact it.

1. **Does the dog enjoy doing whatever it is you're asking them to do?**

   Sadly, a lot of the training we do with dogs is for our egos or our entertainment. We feel good if we can get the dog to perform. We tend to teach tricks that indulge us, rather than life skills. If the dog enjoys what you're teaching them to do and deems it a useful skill, it's likely that you won't need them to be hungry in order for them to co-operate.

   If your dog does not enjoy it, is it really necessary? Is there an alternative?

2. **Does what you're asking them to do cause them pain or discomfort?**

   If so, you probably shouldn't be asking them to do it. We can inadvertently be the cause of our dog's pain, their conformation can make certain acts uncomfortable (like sitting for greyhounds), or they could have an underlying health issue. The environment can also play a part – would you want to sit on cold wet grass in winter? Or a burning hot pavement in summer? Tiles and wooden floors can be slippery and create issues when the dog tries to move around. Compelling them to do something which is causing them pain has the potential to exacerbate any underlying health problems.

3. **Is the dog in a state of stress and unable to focus on the task at hand?**

   Stress can be anything from being overwhelmed by the environment, being in the presence of triggers (other dogs, loud noises, bikes etc.) to chronic stress from a stressful lifestyle, not enough sleep, or the wrong sorts of activities. If your dog is stressed, they are not in a position to learn anything, and the focus should be on stress reduction, not motivation.

4. **Does your dog understand what you're asking them to do?**

   If they don't understand what you're asking, take a look at how you're asking. Is your body language at odds with your request? Do you need to break it down into smaller steps? Is what you're asking them to do sensible from a dog's perspective? One example where we humans often fall down in this regard is asking a dog-reactive dog to 'sit' when they see another dog. First, sitting when you're in a state of agitation goes against everything your body will be telling you to do. Second, sitting is a calming signal, part of a dog's communication toolkit. We'll never understand all of the nuances of canine communication, so expecting your dog to utilise a feature of their communication on command in response to a perceived threat from another dog is rather unfair. Just because he's getting paid for it and therefore complying, doesn't make it feel right.

Remember, if your dog is not interested in taking tasty morsels, something is probably wrong. They are probably stressed or in pain. I'd happily bet that in 90% of situations, it's not a lack of motivation.

CHOICES

Having choices and self-determination is really important for any sentient being. So important, in fact, that one study found that when residents of a nursing home were given three additional choices about their environment, the number of deaths was significantly reduced, compared to a group with no additional choices. Another found that having the ability to exercise control is a key factor in their ability to overcome fear.

In our developed world, it can be difficult to give our pet dogs the freedom they need to simply be themselves. Most cannot roam freely or decide who to interact or not interact with. We expect them to suppress many of their natural instincts – to forage, to defend themselves, to have access to constant social contact. We take away many of their opportunities to make choices about their lives, and this can have a detrimental impact on their welfare.

Having no autonomy is frightening, stressful, and disempowering for any animal. Giving our puppies as much freedom of choice as we reasonably can is a great way to start building a resilient, confident dog. Of course, puppies will need more guidance than an adult dog will, but building independence is essential.

When they are puppies, we can give them the chance to make decisions about some of the things that affect them. Having the opportunity, and the responsibility for doing so, will build their confidence, and encourage them to think rationally and independently. Fostering this independence gives them a better chance of being able to cope when you are not there to make their decisions for them. Imagine how much more frightening it is for a dog to be left home alone when someone else always tells them what to do? Where to sit, what toy to play with, when to eat, when to go to the toilet… and then suddenly they are all alone with no experience of thinking for themselves…

DECISION-MAKING

It is often assumed that dogs are not capable of making decisions, and yet, throughout the millions of years of evolution they underwent before they crossed paths with humans, they have survived and thrived by doing exactly that. And indeed, the 70+% of the world's canine population who are free-ranging continue to make many decisions each day.

And yet, our desire to control our pet dogs means that many dogs live their lives stripped of the ability to make the most basic choices about their own lives – what to eat, where to go, who to spend time with, even when and where to go to the toilet.

The following are some ideas to get you started with choices that you can offer your puppy:

Food: Choosing between different meals, treats, snacks, or chews.

Toys: Have a toy box for your puppy that he can access with his various toys and chews in it. That way he can choose the toy he wants when he wants it!

Movement: Once they are toilet trained, give them the run of the house. They might decide to lie in the sun, or find some nice cold tiles to lie on. They might sleep on the sofa, or go upstairs to find some peace and quiet. Having a dog flap or leaving the back door open for them to go in and out as they please means they have access to more space, and, even better, the toilet. If they 'ask' to go outside by going to the door, pawing it, etc., you can respect this choice by opening it.

Beds: As polyphasic sleepers, dogs will often move between beds a lot. Providing them with access to a variety of beds can help facilitate this natural sleep rhythm. Try providing access to raised sleeping areas (sofas, human beds, or raised dog beds), beds with covers, beds with raised edges that they can rest their heads on, and flat beds they can stretch out on.

Walks: As you begin to bring your puppy out on walks, allow him to investigate at his own pace. Let him choose where to go, and what to do on his walks.

Interactions: Learn to observe your dog's body language. If he seems reluctant to interact with a person or dog, respect that choice. If he seems uncomfortable in a situation, help him get out of it. Imagine how frightening it must be for puppies to be forced to tolerate being touched by every stranger that wants to touch them? Or being forced to say 'hello' to a big, overly friendly dog that they find terrifying? Your puppy will feel much better about the world if they know they can rely on you to back them up and respect their choices when things get scary.

PAWS FOR THOUGHT – CHOICES

Have a think about what other choices are appropriate for your puppy to make, and review this as your dog grows. For instance, while you are toilet training them, you may wish for them to stay in whatever part of the house you're in, but once you are confident that they've gotten the hang of this, you might be happy to give them the run of the house. Try giving your puppy five choices this week!

CASE STUDY: TIRIAN – A PINSCHER PUPPY AND HIS DEVELOPMENT

Tirian was only eight weeks old when he went to his new home, a four-hour drive from his breeder. As we know, eight weeks is too early, ten to twelve weeks old is preferable, but the new owner was aware of this and tried to compensate for it. He slept on her lap the whole drive, and settled quite well at home, spending the night close to his new owner, sleeping soundly.

Indoor puppy mats were in place to make it easier for him to relieve himself quickly, as puppies cannot physically control their bowel and bladder at such a young age.

The first two days were kept quiet and uneventful, and then on the third day he met an adult dog belonging to a family member. Meeting with nice, sensible dogs as soon as possible is important for puppies, as that is what they will miss most when taken from their canine family. People often feel under pressure to keep their puppies away from other dogs until they have completed their vaccines, but if they meet nice, healthy dogs that you know are vaccinated, this is better than keeping them away from other dogs. The absence of other dogs can be really traumatic for them and can cause lots of problems.

On the fourth day, he was carried to a park nearby and let to explore there, and again on the sixth day.

He was not eating very well, so when Turid heard he was on dry food, she suggested the owner make food for him. A puppy's teeth and soft gums can make it a struggle to chew kibble without them getting sore, so it can be better to give them soft food, and, as with ourselves, fresh is always best. In the wild, the adults half digest and regurgitate food for them until they are four to five months old and have more of their adult teeth.

He was up the whole night with diarrhoea – which was no surprise after the stresses of changing home and all the new things he was experiencing. He was then given rice, sweet potatoes, and fish, and got better little by little. A couple of calm days at home probably also helped to get his stress levels down.

On day 11 he had a treat search outside his living quarters and had a dog sitter for an hour while the owner did some errands. Remember, at this age, it is not appropriate to leave your puppy alone without a carer.

During weeks nine and ten he had regular little walks in the park with many things to explore, met different dogs, and also stayed over with a family member for the night.

At 11 weeks old, for the most part he could sleep through the night without needing to go out. He was also eating well and had the opportunity to try a variety of food.

At 12 weeks old he met other dogs daily in the park. He got a slight scratch on his nose from a chihuahua they met, but he did not seem to be bothered by it. The next day he was just as happy to see other dogs! At this stage, he still had the occasional accident indoors, especially during play, which is normal, as their neuromuscular functions are not optimal yet. Make no fuss, just clean up!

At 13 weeks, he was left alone at home for a few minutes, but for a longer period of time had a dog sitter he was comfortable with.

Tirian interacting with an older dog. Courtesy Therese Asp.

Days and weeks went by, and toileting accidents became increasingly rare, he enjoyed meeting many different dogs in the park, and spent a few minutes alone now and then. He regularly did treat searches, and spent time exploring different places and visiting friends.

Tirian still takes a long time to settle down when visiting new places. This is probably due to his stress levels being a little too high, but this might well level out as he grows up and learns to cope with the different activities that were a bit too much for him when he was very young. This could have been avoided by taking it more slowly the first couple of weeks and also taking him home later than eight weeks.

Then, as he was turning five months, and no longer a puppy, he learnt to track, following the track of a person he knew, and finding her. He was very calm and tired afterwards. Using the nose is the best way to lower stress and to teach dogs to concentrate and get into physical and mental balance, so that is advisable.

Puppyhood is over, and new challenges coming as he reaches the young dog stages! He looks like a healthy and fine little fellow, well acquainted with many different dogs. The only training and activities so far have been nose work and exploring, which, together with social training is all they need at this age. Mental stimulation is the most important thing for some time yet.

We wish him a good life after puppyhood!

TAKE-HOME POINTS FOR CHAPTER 3

- Puppies need a lot of sleep; remember that 40% of this should be during the day, so don't skimp on the naps.
- Repetitive exercise should be moderated and your puppy should have opportunities for free movement.
- Stick to the food your puppy was on initially, and do some research into how you want to feed him in the long run.
- Mental stimulation is a great way to use up any excess puppy energy in a calming way – have at least one activity for mental stimulation each day. You can let him do a little nose work, or maybe, as a treat search, finding his teddy bear – anything that engages the nose is good training.
- Give your puppy appropriate choices to build confidence and independence.

4 *Learning life skills*

You may have figured out at this stage that we're not going to be talking you through any obedience lessons – it's simply not the most effective way of having a happy, mutually enjoyable relationship with your puppy.

That's not to say, however, that you don't have lots of teaching and guiding to do over the coming months.

In this chapter, we're going to look at some of the things you can be working on – the useful life skills and good habits that you can begin to form with your puppy.

BRAIN MATTERS…

The brain of the dog is remarkably similar to ours in terms of behavioural and emotional responses.

All of the same basic structures exist in a canine brain as in a human brain. This includes the parts of the brain responsible for learning and memory such as the amygdala, hippocampus, cerebellum and prefrontal cortex. Importantly it also includes the limbic system, the part of the brain responsible for emotions, and research increasingly confirm that dogs have the same rich, emotional lives that we have.

SOCIALISATION AND HABITUATION

In a nutshell, socialisation is the process whereby an animal learns how to recognise and interact with its group members. For a puppy, their parents, littermates, and siblings would all naturally play a part in this. For our pet dogs, socialisation is also likely to extend to their relationship with humans and perhaps other pets. Habituation, on the other hand, refers to the process by which an animal learns about the world they live in and what they need to pay attention to and what they can ignore. When habituation to the environment does not occur, the world becomes a much more stressful place, as the brain has to pay more attention to the environment, treating anything unfamiliar as a potential threat, and triggering more activation of stress pathways in the brain.

Often in the dog world, the term 'socialisation' is used to cover both of these processes.

DOI: 10.1201/9781003305156-4

THINGS TO CONSIDER WHEN SOCIALISING YOUR PUPPY

First experiences are incredibly formative, so the critical thing about socialising your puppy is that their early experiences are positive. If you introduce your puppy to something and they have a terrible, frightening experience, that will stay with them. In fact, research has shown that having just one negative experience with another dog during the first 18 weeks of a puppy's life is enough to cause life-long fear issues with other dogs. This is why it is necessary to ensure that your puppy is in control of this process, and can always move away if anything is too much.

It is also really important not to overdo it. There has been a misunderstanding that puppies need to experience everything within their first 12 to 16 weeks of life (their critical socialisation period), and this can lead to puppies being overwhelmed and stressed, as their humans try to cram far too much into that first couple of months with their puppy. The critical socialisation period is not meant to be one where everything is experienced, just a period when it is wise to start experiencing things, and once they have 'learnt to learn' about new things they will carry this knowledge with them and continue the socialisation process for the rest of their lives.

So yes, they should experience many different things when they are young, but always in small doses that they can cope with. If they learn to cope with these new things, they will be able to apply this learning to future experiences and adapt well to new things and situations. In other words, learning to learn about new things protects them against fear of new things later in life – it is not necessary for them to individually meet each thing they will ever come across.

When

This learning will already be well underway when you get your puppy, if you have gotten him from a reputable breeder. He will already have had nice experiences with people, and will have hopefully become habituated to lots of normal household things. He may have already heard the vacuum cleaner, or learned about stairs. He may have seen people come in hats and carrying umbrellas, or wearing glasses.

How

It's so important to let your puppy take his time when he's encountering the world. Everything will be new, and even when you're not consciously socialising or habituating him to the world, he may be doing it himself! He may want to spend a lot of time hanging back and looking at new things before he approaches them, or he may quite confidently approach new things and give them a good sniff. He may approach something new, and feel a bit overwhelmed, and want to back away and then approach more slowly. Allowing him to take his time and figure it out without too much interference from his humans is the best way for him to learn and become confident.

What

Introduce small shots of some of the things your dog is likely to come across throughout their lives. Below are examples of the sort of thing you might want to introduce your dog to during their puppyhood. Always bear in mind, however, that the crucial thing to take away from this is that the dog must never feel overwhelmed, out of control, or frightened during this process.

- **Humans:** It's great for your puppy to have positive experiences with a variety of different humans in his first few months. Children, adults, older people, people using wheelchairs, people of different races (people often think their dog is racist, but often the problem is that they never met anyone of any race other than their owners' when they were puppies), tall people, bald people, etc. Choose carefully, and select people who you can trust to be gentle and respectful of the puppy. This will help ensure that his experiences with them are positive. If at any point your puppy seems uncomfortable or nervous, make sure he has the opportunity to move away.
- **Clothes and accessories:** Think about the sort of objects humans can have that dogs can find frightening… sunglasses, hats, walking sticks, plastic bags, face masks. You can take a couple of these items, and put them on the ground for the puppy to investigate. Once they have investigated them and seem happy enough, pick one up, and let the puppy see you put it on. He may wish to sniff it, he may just look, he may be a bit nervous. Let him take his time and choose what to do. Of course, he may also be totally unfazed by it! If the puppy is still quite calm and relaxed, you can repeat with another item.
- **Traffic:** A gradual introduction to traffic is important. If you live on a quiet, residential road, where traffic passes reasonably slowly, this is a fine place to start. You can spend some time sitting in the garden with the puppy so he becomes aware of the traffic. Once he is allowed to go out for walks, stick to quiet streets with infrequent traffic. If he is happy with this, you can move onto busier roads for short periods of time, moving back onto a quiet road before he becomes overwhelmed.
- **Other dogs:** It's lovely for puppies to meet up with other puppies for a play. However, it's equally important for them to meet up with sensible, adult dogs to learn how to interact in a more calm way.

Many puppies find traffic scary. Giving them the time and space to observe from a distance can help habituate them to the sights and sounds of traffic. Courtesy Georgia O'Shea.

Going for on-lead walks, and doing calm activities together (such as sniffing and investigating) will help your puppy develop and maintain his canine communication skills.

- **Other animals:** This could be particularly significant if you live in the countryside! Again, it's important these experiences are positive for your puppy, so allowing him to be chased by a herd of sheep is not going to be helpful. Allowing him to investigate other animals from a safe distance, smell where they've been, or meet through fences is a much better way of ensuring your dog can peacefully co-exist with them.

Scout observes some ponies from a safe distance. Courtesy Georgia O'Shea.

Remember, when it comes to socialisation, a good puppy class can be helpful with the socialisation process, but a bad one can be catastrophic. See more about choosing a good puppy class on page 76.

The movement, sound, and reflection from the television can take some getting used to! Georgia O'Shea.

CASE STUDY: HABITUATION – ANOTHER CAUTIONARY TALE

When Steph started working as a dog trainer in London, one of her first clients was a lovely little labradoodle. He was a very sweet-natured, gentle chap, and a little on the nervous side. Steph spoke to his humans about socialisation and habituation, and the need to introduce new things slowly, and in a way that the puppy could cope with it.

Obviously, she was unclear. The next time she came to see them, they told her they had brought the puppy to the carnival in central London, where there were crowded streets, loud music, and mounted policemen. They thought it was a great way to get him used to things. The puppy was petrified, not habituated. He would probably remain terrified of all of these things for a long time!

FORMING GOOD HABITS

'Life is no obedience lesson, it's a way of living together' – *Turid Rugaas*

When young individuals grow up, whether children or puppies, they need time to learn how to behave as adults. Being obedient or well-behaved is not the most significant thing at this stage, nor is learning lots – young children are not supposed to take university exams, and puppies are not meant to be model dogs. They are still under construction, and the best things to focus on teaching them when they are young are some good habits. The brain is not fully developed yet, and can only learn a little at a time. If we try to teach them more than they can handle, they will only become stressed and frustrated. Learning a few habits means that they do some things automatically, and no longer need to think so much about it, leaving a little more space for learning.

A habit is learned by repeating something consistently, and when teaching a new habit, we can help the learner in the process so they can get it right.

For instance, we teach children to put on clothes, tie shoelaces, brush teeth, and so on, and once they have learned to do it, they do it without having to think about it. Less bother for us, less brain work for the child!

Puppies can also learn a few habits to make life easier for them and for us. Teach one at a time. It usually takes two to three weeks for a new habit to become established, so a dog can develop many useful habits during their first year!

Good habits to teach your puppy include the following.

NOT TO JUMP ON PEOPLE

Read more about how to deal with this in the common problems section.

TO PAY ATTENTION TO THE CLICKING/SMACKING SOUND

It is really useful to teach your dog to respond to a clicky, kissy, or smacky sound (whatever sound you can comfortably make with your mouth!). This is simply a noise to get the dog's attention.

How

To teach this, make the noise, and immediately give your dog a treat. Repeat a few times. When you think your dog has made the association between the noise and the treat, make the noise, walk away a couple of steps, and give the puppy a treat when he follows.

When

You can use your attention sound to let your dog know that you're going to change direction, to ask them to follow you, or to ask them to look or move away from something.

TO WALK ON A LOOSE LEASH

We don't recommend you teach your dog to 'heel' in the conventional sense, as this really restricts their freedom on their walk. Rather, teaching them to keep their lead nice and loose means you can keep your arm within your shoulder joint, and they can potter about and sniff things, and curve if they need to.

How

- First of all, teach your puppy the attention sound. You can practice this on- or off-lead. The best place to practice is somewhere that you have lots of space.
- Make the sound (a click or kissy noise with your mouth), and when the puppy pays attention, turn your body. He will follow the way your body is facing.
- Make the sound and turn again, moving a few steps when he follows.
- Continue like this, making the noise, and then walking a little in the opposite direction when the puppy follows.
- Now you can gradually move further before changing directions, etc. The puppy can potter about and sniff, and when you make your noise, the puppy will understand that it's time to follow you.
- Put on the lead and repeat. If the dog pulls at any point, or is about to pull, turn in the opposite direction, and when he looks at you, make your noise and walk in another direction. Do not pull your dog, keep the lead loose and wait for him to follow.

LOOSE-LEAD WALKING – THINGS TO REMEMBER

- Use a harness and a long enough lead when doing this exercise. If your lead is too short, most of its length will be taken up going from your hand to the dog's harness, and they will have no space to move around.
- Really slow down – you'll probably find if you slow down, your puppy will too.
- Keep training sessions nice and short. Puppies have very short attention spans!

Always face the way you want your dog to move – do not stand facing them, as, to the dog, this body language is signalling that you don't want them to come closer.

Read more about teaching your dog to walk on a loose lead in Turid's book *My Dog Pulls, What Can I Do?*

TO RELAX WHEN WE SIT DOWN

How

Have some beds for the puppy near your chair. Put your puppy's lead and harness on, hold on to his (longish) lead, sit down in your chair quietly and just wait for your puppy to settle. Do not give him any instructions or ask him to sit. Do not tell him off if he pulls away or barks. Just let him take his time and decide of his own accord to relax. Once he does settle down, don't praise him or pet him. The feeling of relaxing will be reward enough. Just remain perfectly still and silent for five or ten minutes and let the puppy relax.

When

You can practice this a couple of times a week. Initially, choose a time when your puppy is already reasonably calm, rather than when he is having a burst of energy and tearing around the house! Start somewhere that's not too distracting, such as your house or garden. As your puppy gets the hang of things you can begin to gradually move to more distracting areas – maybe a quiet area of the park, perhaps a friend's house when there's not too much going on, then on to a coffee shop at a quiet time, etc.

Why

Your puppy will learn that when you sit down, he can sit down and relax too. This is a great skill if you ever want to bring your puppy to the office or to a coffee shop or pub.

From a young age, Scout got used to spending calm time on boats with his humans. Courtesy Georgia O'Shea.

Scout relaxes while his people visit a bike shop. Courtesy Georgia O'Shea.

Loose lead walking – allowing the walk to be pleasurable for both dogs and humans. Courtesy Nina Ozmec.

TO ELIMINATE OUTSIDE

Read more in the section on toilet training.

TO ACCEPT BEING GROOMED/TO STAND STILL FOR *SAUMFARING*

Start when the dog is close, and you're sitting still (this is less intimidating than bending over your puppy). When he wants to leave, let him. Just sit still waiting for him, and when he comes back, do a little more. Remember to be really gentle, and to be calm and friendly yourself. In the beginning, maybe scratch or stroke him a little, and do a little bit of brushing. Holding the puppy tightly and being insistent is the worst thing you can do – all living creatures feel like escaping when they are held, so do avoid this. Allow your puppy to take his time, and he will get to the point where he will be happy to stand for long enough for you to do what you need to. Short sessions and ensuring your puppy can always choose to move away is the best way to make it a pleasant experience for life.

TO RESPECT THE HAND SIGNAL

The universal hand signal is one of the most helpful and efficient tools in daily life. All mammals – and even birds – respond to it. It is something they understand innately so you can use it without any 'training'. If you do not use anything else with your dog, this signal is what we would recommend using. It will be a great help as the dog grows up and in many situations that you will encounter.

Respecting the hand signal.

A relaxed showing of the palm of your hand, and the dog will stop trying to get on the table, reacting to something outside the window, feeling unsure about something scary, pestering you to let him out 50 times in an evening, and many other situations that will undoubtedly pop up during his life.

Work has been done examining the effect of this hand signal on dogs' heart rates. The results show that with no prior training or experience of the hand signal, a dog's heart rate remains lower when their human moves away from them after first showing them the hand signal, compared to when the human moves away from them without showing the hand signal. A lower heart rate generally correlates with lower stress levels, while a temporarily increased one can indicate higher levels of stress, anxiety, or other strong emotions.

It is the calmest, most gentle and efficient thing you can ever use, so do practice before he gets into the teenage period and might be a little deaf – the hand signal will help you through that period too!

TO RETURN TO YOU WHEN CALLED

The best way to teach a dog to come when called is to start when he is a puppy and wishes to be near you.

Several times a day, when the puppy is awake and pottering around, just make a sound – like clapping your hands, or a clicking sound – and take a few quick, very short steps away from the puppy. They cannot resist the movement and will follow you!

Say 'come' the moment he gets to you, and praise him in a happy voice.

As time goes by, you can sometimes run a little further away – maybe eight or ten steps, or into the next room – but do not hide. That can scare a puppy or cause him to become frustrated, and we do not want that. We want him to associate coming when you call with happiness!

As he grows up this will simply be another good habit, and a happy one.

Learning to come when called! Courtesy Turid Sunde.

HABITS THAT WON'T FEEL GOOD

Good habits should feel natural. Unfortunately, habits can also be taught by force and by commanding the dog to do things, but no matter how much food you use or how positive your approach is, the odds are that the dog will never perform this sort of habit with any pleasure and might well feel bad every time he does it. This has no value and will not deepen your relationship with your dog. Feeling bad creates stress and negative reactions, and often these habits are totally unnecessary anyway.

The following are some examples of learned habits that your puppy will not feel good about.

Going into a crate

Being locked in a cage is contrary to any animal's instincts. All animals want to move freely and exercise choice. A puppy may learn to accept being in a crate if we give them no other choice, but often this is because they have realised that resistance is futile, and they are in a state of learned helplessness. This is not a healthy state of mind for your puppy.

Sit-stay

As we discuss on page 81, sitting is not a common resting position for our dogs. Compelling them to sit and stay while we move away from them is stressful, and often uncomfortable. The more stressful the environment is, the more stressful this is for a dog.

Making eye contact on command
Direct eye contact is quite confrontational in the dog world, and polite interactions for dogs involve lots of looking away, avoiding prolonged eye-contact, and frequent blinking. Asking our dogs to make eye contact on command is therefore contrary to their own communication instincts. Additionally, craning their necks to look up at us is physically uncomfortable, and if we ask them to do this while moving, will pull their bodies out of alignment.

Walking to heel on your left side
Our dogs' walks should be one of the highlights of their day – and for dogs that means investigating the world around them with their noses! It is perfectly normal for our dogs to want to sniff lampposts, walls, hedges, parked cars, and anything else they encounter, and if we insist upon them walking to heel on one side of us, they can do none of this. What's more, your dog or puppy may wish to curve to avoid a direct approach if they spot another dog or person, and insisting they make a straight-on approach will be stressful for them. Moving naturally should be part of the joy of the daily walk, so instead of heel-walking, work on some loose-lead walking, so both you and your dog can enjoy the walk.

If your dog has many good habits as he grows up, he will most likely not need to learn many commands for control. Maybe none at all. His behaviour will be pleasant in daily life, and you will not feel the need to order him to stay, sit, lie down or any of the other traditional commands that dogs are taught. You will have a well-behaved and very pleasant dog, without undermining your relationship with him.

So, why not make yourself the good habit of teaching your dog good habits?

PAWS FOR THOUGHT – GOOD HABITS

Make a list of the habits you're going to teach your puppy over their first year. Put it in your diary and review it periodically to see how you're doing.

CHOOSING A PUPPY CLASS

A well-run puppy class can equip both you and your puppy with great skills. However, if you have access to nice adult dogs who your puppy can socialise with, it is certainly not a necessity – it is much better to swerve a puppy class altogether than to end up in a bad one! If you do decide to attend puppy classes, it's really important to choose a class and a trainer that will stand you both in good stead. Here are some things that we suggest you look out for:

Ethos: Today, we talk more about small social groups for puppies, rather than actual classes. Puppies should not learn obedience during their puppyhood – the exercises are too hard on their developing musculoskeletal system and too much for their very short attention span. They should also have just a few other puppies and humans to deal with at a time, as they get easily overwhelmed and stressed by too much going on. Small groups of two to four puppies of much the same age and size (to avoid bullying as much as possible) with a focus on learning social skills, is the ideal set-up at this age. They can learn a few exercises such as learning to walk on a loose lead, while owners can learn how to care for their puppy and provide mental stimulation and how to handle things like grooming, how to stop the puppy jumping up on people, etc. We are hopeful that there will be more groups like this in the future and believe that dogs will really benefit from a move away from a focus on obedience and control.

Trainer: Find out about the trainer. In a lot of countries, dog training is not regulated and anyone can set themselves up as a dog trainer. Find out about the training the trainer has received and if they keep up with continuing professional development. Check if they are members of any professional bodies and find out if the ethos of their accrediting body meets with your approval. You will probably be paying a lot of money for your puppy classes, so make sure you're getting a knowledgeable trainer using humane, up-to-date methods!

A small, calm puppy class.

Methods: Ask if you can go along and observe a class. Steer clear of any trainers using force, fear, or pain to train puppies. Puppies in slip leads or choke chains or being pushed or dragged into 'sits' or 'downs' are both bad signs.

Size: Smaller numbers mean more attention for each puppy, and more scope to deal with the particular concerns of the group. We suggest limiting the puppy class to four puppies and find this to be a good number to ensure that everyone has enough space, puppies can be given plenty of freedom and individual time, and are less likely to be overwhelmed. Some trainers will take six or eight puppies but may have a lot of space and an extra trainer along to help out. Ask how many puppies are accepted onto the course you're looking into, and look at the space available.

Recommendations: Speak to people who have attended the classes. Don't just ask them how they found it – many people have nothing to compare their puppy classes to – ask for specifics. Find out if they felt they learned about what puppies need to grow into healthy happy dogs, if they had time to ask questions about issues they were experiencing, and if they felt supported by their trainers. Ask about what methods they used, and what was covered in classes.

What's covered: Many puppy classes teach tricks rather than life skills. While it may sound like fun to teach your puppy to sit, roll over, and give the paw, wouldn't you rather learn how to raise a happy, confident puppy who can cope with life? Some classes will also equip you with knowledge about mental stimulation and how to teach your puppy things you can enjoy doing together, like scent work.

Once you've chosen your puppy class, remember that you still need to be your puppy's advocate. If you think your puppy is not enjoying the class, is feeling overwhelmed, or is being bullied, speak up, help him out, and if needs be, leave. It's very easy to fall into the trap of doing what you're told by a person in a position of authority. But if it doesn't feel right, it probably isn't right.

PUNISHMENT – AND WHY IT SHOULD BE AVOIDED!

Without getting into behavioural definitions, punishment in training can include anything that hurts, threatens, or deliberately frightens your puppy. When you are embarking on your journey with your new puppy, it is vital to have an understanding of the consequences, both longer- and shorter-term, of using punishment in dog training. Many people will offer you unsolicited advice about training your puppy, and some of these may advise using some form of punishment; if you are having a bad day, or feeling at your wits' end with an infuriating puppy, you might just be tempted. Having an understanding of the pitfalls of punishment might help you resist the temptation! However, before you read on, take a moment to forgive yourself for mistakes you might have made with previous dogs, when you knew less about canine behaviour. We have all done things with our dogs that we would have done differently or not at all with the benefit of hindsight.

There are many issues with using punishment as a training tool. It causes the dog pain and creates fear, it stifles curiosity and the desire to learn, it undermines our relationship with them and can cause dogs to completely shut down and live in a state of learned helplessness – a major welfare concern. However, one of the most alarming problems that can arise for owners using punishment in training is the increase in aggression that can stem from it, both as an immediate response and as a more long-term effect.

In 2009, a veterinary behaviourist in the United States performed a study assessing the impact of various training methods, including physical corrections (e.g. hitting) on dogs' behaviour. Her findings were that dogs who were subjected to confrontational training methods very often reacted aggressively (see Table 4.1). In contrast, it was found that reward-based training elicited aggression in very few dogs, regardless of the training issue.

Apart from the more immediate risk of an aggressive response to punitive training, there also appears to be a more long-term effect on the dog's behaviour. A further study in 2010 found that dogs subjected to physical punishment were more aggressive than those who weren't.

Other research, as well as reiterating the findings of the above studies, found a link between owner use of punishment and an increased level of reactivity towards dogs from outside the household.

This is just a sample of the many studies out there implicating punitive training as a risk factor for aggressive behaviour. So much so that, in light of the evidence linking punishment-based training and aggression, it has been argued that punishment-based dog training is a public health risk factor.

Table 4.1 Aggressive Reaction to Confrontational Training Methods

Confrontational training method	Frequency of aggressive reaction
Hitting or kicking	43%
Growling at dog	41%
'Alpha rolling'	31%
Staring at/staring down	20%
Forcing into a 'down'	29%
Grabbing by jowls and shaking	26%
Shouting 'no!'	15%

As well as having a greater chance of inciting aggressive behaviour in your dog, punishment-based training methods are less effective than their reward-based counterparts.

Studies have found the following:

- Dogs who were trained using reward-based methods had the lowest occurrence of over-excitement.
- Dogs who were trained using punishment-based methods (including verbal reprimands) had the highest occurrence of separation-related problems.
- In none of the tasks trained were punishment-based methods more effective than reward-based methods.
- Dogs trained using exclusively reward-based methods were reported to be significantly more obedient than those trained using either punishment or a combination of reward and punishment.

They also found a positive correlation between the number of problematic behaviours reported and the number of times owners reported using punishment-based training methods.

Additionally, both fear and pain trigger the 'fight or flight' (stress) response in our bodies. Occasional stress is something we can cope with, and can be necessary to our survival. Chronic stress, on the other hand, is disruptive and harmful and has physiological, mental, and behavioural effects. Subjecting our dogs to chronic stress through the use of punishment can be considered another way of inflicting suffering, and this is contrary to the duty of care we have to the animals we share our lives with.

So, if punishment-based training is less effective than reward-based methods, has an adverse effect on a dog's health by causing them stress, and is associated with an increased likelihood of eliciting an aggressive response, why do these methods remain in use?

Often, when we shout at a dog, hit them, or jerk the leash, the behaviour is interrupted. For instance, the dog is barking at another dog and we shout and he stops because he is startled and perhaps frightened. The fact that the owner will probably feel better for having vented their anger, and the fact that the behaviour stops in the instant, make the act of punishing very rewarding for the punisher. But chances are, the next time a dog passes he will bark again. In fact, there's a good chance the behaviour will escalate, as the dog is likely to associate the other dog's presence with the owner's unpleasant behaviour. And as the evidence illustrates, this is not an effective tool for long-term behaviour change.

Most people do not want to cause their puppies or dogs pain or fear, undermine the relationship they have with their dogs, or run the risk of their dog developing aggressive tendencies. Rather, it seems that punishment is something people often resort to when they don't know what else to do. It is our job to give you the skills and understanding you need to raise a puppy without resorting to punishment!

WHEN SHOULD PUPPIES LEARN TO SIT?

The answer is very easy: they should not. They can of course sit when they feel like it, but it should always be their choice.

WHY?

There are several reasons for this.

When puppies are born, their development is not yet complete. They continue to develop physically: the neuromuscular system develops so they can control their bladder. The balance system in the inner ear continues to develop, the brain is the last to be complete with prefrontal cortex, and, crucially, the joints are not properly connected yet. X-rays have revealed that at birth, a puppy's joints are not connected at all, and instead, are floating around in the sockets. That is why they cannot stand or walk. The joints are still not well connected when you get your puppy, and making them do exercises that make the joints slide around without proper support can create life-long damage. The joints must have time to develop properly before we can demand anything physical of our puppies.

Secondly, they have not yet developed sufficient muscle either. Muscle development is required to support the joints and indeed the whole body, and to facilitate movement. Building muscles takes time, and they must be built from the inside out – all of the deep muscles that are so important for locomotion and balance must develop first. This is best done by free movements, by choice, to give the variety that is necessary, and so that the puppy can decide when they've had enough. It takes the greater part of a year for a puppy to build the muscle necessary for good balance, movement and coordination.

Additionally, some breeds are bred for specific jobs, which can affect their physiology – for instance, sighthounds are structured in such a way that makes it hard for them to sit, with their long thighs. They suffer when the hind legs are compressed in a sit. They rarely sit by themselves – some of them never sit if they can avoid it.

Puppies of heavy breeds will need more muscles than others to sit down and get up again, and because they do not have much muscle as puppies, they have to jeopardise their joints to make these movements, creating a lot of discomfort. Never ask a heavy breed puppy to sit!

Observing all these things over the years, Turid started to ask her dog-training students to do studies on how often, when, and how dogs actually sat without being prompted. The study is ongoing, but they have already observed thousands of dogs and had some amazing results.

Firstly, dogs sit a lot less by choice than we make them sit. They prefer standing or lying down. There are particular situations in which dogs might sit, and that was a surprise.

REASON 1

When they need to look at something from a distance, they must raise the head and curve the neck to be able to see, due to the construction of the eyes. After a few seconds, it starts to ache in the neck muscles, so they sit down to level out the curve, and then they can sit watching as long as they want.

This was overwhelmingly the most common reason for the dogs studied to sit.

REASON 2

Sometimes dogs sit briefly on the way down to, or up from, a down position. This is usually due to a lack of muscles preventing them from getting straight up or down. That is seen mostly in puppies, old dogs, and dogs with physical problems, to make the transition easier.

REASON 3

A learnt sit.

Dogs who had been drilled to sit in all kinds of situations did it as a habit. When people stopped asking them to sit, it would gradually get less and less.

Further down the list, we found some reasons such as sitting down to calm somebody, sitting because a lack of balance made it hard to stand, and some unidentified situations in which it appeared that the dog had pain standing.

Beware of ordering your puppy to sit. It may cause them discomfort. Courtesy Lise Rovsing.

In other words, it is not a good idea to ask puppies to sit before they are fully developed and have the balance, coordination, and muscles that might make it easier.

Observe your puppy and see how often he sits by himself, and in which situations. Also, note the way he is sitting. Sitting with one or both hind legs stretched forward should tell you that it is uncomfortable for him to smash the leg joints together in a sit. Maybe he sits on one 'ham', on the side, which is also to release the pain of the joints in a sit.

You can learn much by observing and learning what is best for your puppy. He will know best what feels good and bad.

SOME KEY POINTS FOR POSITIVE BEHAVIOUR CHANGE

If your puppy is displaying a behaviour that you want to change, you may need the help of a good trainer or behaviourist, and the exact plan of action will vary depending on the problem. However, below are some key things that should be helpful to bear in mind regardless of what the issue is.

- Your first step is always to take away the opportunity to practice the behaviour. This is important because the more the behaviour happens, the better your puppy becomes at it, and the more likely it is to happen again. So if, for example, your puppy is running away from you in the park to play with other dogs, you will have to keep him on a long lead until you have changed the response. If you are waiting for advice from a trainer or behaviourist, removing opportunities for the behaviour to happen is one of the best things you can do while waiting.
- Try to identify the cause of the behaviour rather than focusing on simply changing the behaviour. Does your puppy have a need that's not being met? Is something frightening or worrying them? Could they be in any sort of pain or discomfort?
- A stressed puppy or dog is not in a position to learn anything. Always spend some time removing stressors and making sure you're meeting all of your puppy's needs as laid out in this book (that they're getting enough, good-quality sleep, are eating appropriately, not being left alone or allowed to regularly become distressed or excited) before trying to embark on any sort of behaviour change.
- Make sure the behaviour is not being inadvertently rewarded. If your puppy is jumping up and you're looking at him, pushing him down and telling him 'no' or 'down', you're giving an awful lot of attention and feedback for a behaviour you don't want to encourage!
- Always work within your puppy's comfort zone – if you are trying to train your puppy not to bark at the cat across the road, there is no point trying to work on the behaviour when he is displaying it. Start your work when your puppy is far enough away not to be reacting and build from there.

TAKE-HOME POINTS FOR CHAPTER 4

- Puppies don't need lots of training – they just need to learn about life and develop some good life skills.
- Begin to introduce your puppy to the world, but don't overwhelm them – the key thing about socialisation is that it should be positive and manageable. They don't need to experience everything in their first few months.
- If you're opting for a puppy class, choose wisely!
- Avoid punishment – there are more effective and safer ways to train your puppy.

5 *Common concerns for puppy owners*

NIPPING

We have so much good news for you about nipping! First of all, don't worry! No amount of puppy nipping is an indicator of an 'aggressive' dog.

Although it may not feel like it when your puppy is merrily biting your hand with his sharp little teeth, nipping is a perfectly normal, natural part of being a puppy.

There is an eminently sensible evolutionary reason for this behaviour. Dogs are a social species, and once fully grown their jaws are potentially a very powerful weapon. As puppies, they need to learn to control their bite and to learn what is a hard, dangerous bite, what is a gentle 'reprimanding' bite, and what is a soft play bite. This means that when they are fully grown, with powerful jaws, they won't accidentally harm a friend while playing, and that a warning bite won't do unintended serious damage. While puppy teeth are sharp, they are generally not capable of doing much damage, so while they have these baby teeth, they can safely experiment with their bite.

Remember, in the wild, your puppy would be with their mum and their siblings during this period. If he bit his mum too hard, which happens when they start getting teeth but have not yet learned to use them properly, she would quickly let him know by squeaking. This would surprise the puppy and he would stop right away. It can be effective if we squeak when they bite whilst they are still very young. Later on, this is a less effective technique and can simply wind them up further.

It is also important to remember that our skin is much more sensitive to biting than the fur-covered skin of a puppy's mother and siblings, so the bite often feels harder to us than the puppy intends.

Puppies also use their mouths to investigate the world – they pick things up, taste them, give them a chew, and decide if they're of any interest! Flowing items of clothing such as dressing gowns and pyjama ends are often particularly tempting, and when puppies are greeting family members in the morning, these items will often fall victim to puppy teeth!

Another reason a puppy might be 'mouthy' is because they are teething. As with humans, puppies lose their first set of teeth and grow another, bigger set, to do for life. This is a painful process and during this time you may find they embark on slightly more destructive chewing of items. Ensuring they have plenty of appropriate things to chew can help keep your chair legs and remote controls safe!

The second piece of good news is that most puppies naturally grow out of this behaviour when their big teeth come in.

Bearing all of this in mind, there are a number of things you can do to help reduce the frequency and intensity of puppy-nipping if you are finding it troublesome.

DOI: 10.1201/9781003305156-5

A TOOTHY TIMELINE

- **5–6 weeks:** Your dog's baby teeth have come in – 28 in total.
- **3–4 months:** Your dog's baby teeth will fall out and his adult teeth will begin to come in. You will often see a little bit of blood on chews or toys that they have been chewing, and might be lucky enough to find the odd tooth!
- **6–8 months:** Your dog should have all 42 of his adult teeth. The puppy-nipping should be a thing of the past by now.
- **10–12 months:** The roots of the bigger teeth finally finish embedding into the jaw bone – you may get another bout of destructive chewing around this time!

Checking out a cushion with his mouth. Courtesy Alja Willenpart.

WHAT TO DO

First of all, take a look at your puppy's state of mind – this can affect how nippy he is. Ask yourself the following questions.

Is he tired?

As we have seen, puppies need a large amount of sleep, and more often than not, they are not getting quite enough of it. If your puppy is being particularly nippy, try sitting down in a quiet room with the puppy and staying put while he has a snooze. Giving him an appropriate food-based chew can help interrupt the nipping and settle him down for a rest.

Is he over-stimulated?

The world is full of new things for your puppy, and for most puppies investigating their new environment is more than enough stimulation for them. Too much fast, exciting play will wind the puppy up and leave them full of adrenaline, and this can lead to more nipping. Have they had a few busy days in a row with lots happening, new people, or new places?

Is he frustrated?

A major cause of frustration in puppies is crate-use. Once they are released from the crate, they often vent their frustration at having been confined in a boring cage on their own by engaging in nipping and mouthing. Puppies can also become frustrated by constant handling. This is more common in households with young children, but grown-ups can be just as bad! If you are holding your puppy or playing with him, and he begins to nip, take a step back and give him some space.

Is he contact seeking?

On the other hand, puppies also use nipping as a way of seeking physical contact, as they do not have many other ways of seeking it. If your puppy feels a little under the weather or lonely, or in need of attention (and as with human babies, they need a lot of attention!) this too can lead to nipping.

SCOUT'S DIARY – DAY 18

'He is still very excited in the morning to greet everyone but has given up pulling on dressing gowns, we didn't make a big deal out of it and in not reacting we hoped the fun would wear off... and it seems to have!'

An enthusiastic morning greeting from Scout. Courtesy Georgia O'Shea.

PRACTICALLY SPEAKING...

Make sure your puppy has things to chew. Puppies need to chew, and ideally he should have easy access to a variety of items that can satisfy that need – food-based chews, rope toys, wooden chews, soft toys etc. Beware of just producing these items when the puppy is nipping you, or he'll quickly learn that human-chewing is a great way to solicit something tasty. Having them in boxes or baskets so that the puppy can choose the item that most appeals is a good idea.

If you can, pre-empt the behaviour. Does your puppy get very nippy in the morning or evening? Or when the children come home from school? Give him a chew just before the behaviour typically begins.

Use your body

Standing up, turning your back and looking away, will be understood by your puppy as a clear message that you do not like the behaviour. Using the hand signal (see page 74) while you do this ensures that the puppy feels acknowledged rather than ignored.

Move away

If the puppy is really engrossed in his nipping and turning away doesn't help, it can be helpful to step behind a barrier. Puppy gates are great for this, as you can prevent the behaviour without the puppy feeling isolated. It is best not to put the puppy behind a door on his own, as this can increase his stress and frustration and lead to an increase in the behaviour once he is reunited with his humans.

WHAT NOT TO DO

Do not punish the dog, hold his mouth shut, lock him away, or shout at him. This is normal puppy behaviour. Over the next month or so, he is learning bite inhibition from the reactions he gets to the pressure he applies with his mouth.

Definitely don't make a big deal of it (remember, attention is the number one reinforcer for most dogs!). In most cases, it will have greatly reduced of its own accord by the time the puppy is about five months old, and in the meantime, a bite from a puppy is generally not dangerous.

Where the nipping behaviour is particularly pronounced or continuous, it is nearly always a case of the puppy being over-excited, over-tired, or frustrated, so do take a good look at the puppy's overall routine and see if there's anything you can do to tone things down and ensure his needs are being met in a species-appropriate way.

CASE STUDY: NIPPING

When Steph was running puppy classes in London, she had a lovely little ten-week old miniature schnauzer join. He lived with three generations of women – a girl of about ten, her mother, and her grandmother. He was their first dog, and the grandmother had always been a bit nervous about dogs, but this sweet little guy was winning her over, and they were getting on great. They finished the six-week course, but several weeks later Steph got a call from the mother...

The puppy (now about five and a half months, and reaching the age where you would expect the nipping to be reducing) was terrorising the grandmother. He was hiding under the table, biting her feet, and because she was not that mobile, she was struggling to move away from him. She was getting progressively more frightened of him, and they were at a loss as to what to do, worried that that dog was becoming aggressive.

Steph went to see them in their home, and it became immediately clear what the problem was. Each time the puppy bit her foot (which was appealingly clad in a sheepskin slipper!), the grandmother would say 'oh!' and pull her foot away. He would then bite her other foot, and she would say 'oh!' and pull that foot away... the puppy thought this was a terrific game! And because the granny wasn't removing herself or ignoring him, he probably thought she also loved the game.

In other words, he was being rewarded by the interaction and attention he was receiving, so he continued the behaviour. Luckily, this was easily solved. The grandmother repeatedly moving away or getting up to turn her back each time he bit her was not feasible. So instead, Steph suggested they put a lowish barrier on one side of the table that blocked his access. Each time he bit her foot, she would gently pick him up, and put him on the other side of the barrier. This removed the opportunity for him to be rewarded for the behaviour, and when he was no longer getting a game out of biting feet, he soon got bored of it.

Steph has also commonly come across cases where the puppy continues to nip the child or children of the family for some time after he has stopped nipping the adults. There can be a number of reasons for this, but a common cause is that the children's reaction to being nipped is perceived as being more fun or engaging to the dog – similarly to the grandmother in the case above, they will jerk their limbs away, make exciting noises, or try to run away, and the puppy thinks it's all part of the game!

TOILET TRAINING

Toilet training is one of the first things to concern many new puppy parents. Significant variety exists in how long it takes to train any given puppy, and this can cause great anxiety for humans. It may be that this puppy is taking a lot longer to train than your last puppy, or that your friend's puppy was going outside within days of coming home, while you're still mopping up rivers of wee six weeks later. There are many factors affecting how long it will take to toilet train your puppy, including:

- The age they are when you get them.
- The work the breeder put into toilet training.
- The breed of dog (for example, French bulldogs and pugs tend to be a bit slower in this regard!).
- The dog's temperament.
- The time of year (lots of puppies don't like going out in the cold and rain and finding a nice, quiet warm corner to do their business in is much more appealing than heading outside! Summer puppies are often quicker to train for this reason).

The good news is that dogs are naturally clean creatures, and if we did nothing at all, they would eventually toilet train themselves. They have an instinct to be clean where they live, and, over time, learn that our entire house is their home. However, as a puppy, they have many factors working against them – they do not have the neuromuscular ability to choose when and where to go to the toilet until eight and a half weeks, they cannot hold their bladder or bowel for long, and they have the attention span of a gnat. So figuring out in enough time that they need to go, and then getting themselves to an appropriate spot, is an unfair expectation to have of your puppy initially!

From eight and a half weeks old, they begin to gain a little more control, and as soon as they can physically control their elimination, they will try to get to the right place to go. All you need to do is make sure they can access appropriate areas! How long it takes for them to be fully house-trained depends largely on how effective we humans are at providing enough opportunities for them to get it right.

By five months old, however, most puppies will be toilet trained. And while this may sound like an eternity, bear in mind that it takes over two years to toilet train a human child!

That's not to say there aren't things you can do to try and speed up the process, however…

Success! This puppy pees outside on the grass. Courtesy Jolly Doodles.

WHAT TO DO

1. Give your puppy plenty of opportunities to get it right. Bring him outside regularly, especially first thing in the morning, after food or water, after any play or excitement, and before bed.
2. As soon as he goes, reward him with very tasty treats! You only have a very limited period (two to three seconds!) within which he'll associate the toileting and the treat, so be prepared! Be careful not to start your praise too soon, or you might distract him from his business.
3. Ignore any accidents and clean thoroughly with enzyme-removing products. After any accidents, ask yourself if there were any signs that the dog wanted to go out that you might have missed – sniffing the ground, pacing, circling.
4. Scatter-feeding your puppy on the floor in rooms where he has accidents can help deter him – no one wants to pee where they eat!

It is also worth noting that tiredness and over-stimulation are also your enemies when it comes to toilet training. If your dog is tired or wound up, they will not have the focus to think about needing to go to the toilet and getting to the right spot.

People often forget that play and excitement will trigger a stress response in your dog, which makes them need to empty their bladder. This is why they will often have an accident just after someone comes home or when a guest arrives.

If your puppy was living with their mother in the wild, she would simply clean up any poo in the den by eating it! This helps deter predators who might otherwise smell the faeces and be attracted to the den, as well as keeping the area clean.

She does not reprimand the puppies, and your puppy will not understand any anger you might feel about this.

WATCH OUT FOR CLUES YOUR PUPPY MIGHT NEED THE TOILET:

- He is circling.
- He is pacing.
- He tries to sneak off somewhere.
- He is sniffing the ground.

WHAT NOT TO DO

It is best not to tell the dog off for accidents. This includes startling him by interrupting him, picking him up, shouting, gasping etc. Certainly don't punish him! Dogs who are punished or told off for toileting indoors are liable to refuse to go to the toilet on walks, because they will associate going to the toilet in front of you with being in trouble.

A NOTE ON USING CRATES TO TOILET TRAIN

We would also caution against the use of crates for toilet training purposes. As well as the ethical considerations involved in locking a dog in a cage, their use as a toilet training aid can backfire. As mentioned above, dogs are naturally clean creatures and will want to move away from their sleeping area to toilet. The idea of using a crate to toilet train is that the dog will be forced to either hold their bladder or soil their sleeping area. If your dog gets to the point when they simply cannot hold their bladder/bowl any longer (an extremely distressing experience, and one that is not good for your puppy's health) and are forced to soil their sleeping area, they can begin to override the instinct they have to be clean. If this happens, you can face a much greater toilet training challenge in the long run.

> **A TIP FROM TURID**
>
> When the puppy has done his business outside, try not to startle him with praise and treats. This can distract him from the real intrinsic reward – the feeling of relief at having emptied his bowel or bladder.
>
> As we mentioned, dogs would toilet train themselves if we did nothing, so feel free to simply give them the opportunity to teach themselves.

TROUBLESHOOTING

In our work with puppies, we have noticed some common issues that people encounter whilst toilet training their puppy. Some of these are outlined below.

- **The dog is not going to the toilet on walks at all but is relieving themselves as soon as they get home.**

  A common complaint, which seems to be linked to confidence or focus. There is so much going on outside that can be a bit scary, or just really interesting, that the puppy can't relax enough to go to the toilet. This tends to resolve with age and experience. But in the meantime, bringing the dog straight out to the back garden once they get home from their walks can manage the problem.

- **The puppy is peeing on his walks, but not pooing.**

  Another very common complaint. This often appears to be a result of nervousness. It takes longer than peeing and the dog is in a more vulnerable position, so he feels safer doing this at home. As the dog becomes older and more confident, this generally resolves. Keeping walks to quiet areas with fewer distractions can help.

- **Getting it right when you're around but having accidents when you're not.**

  This can be a case of the dog being left too long and simply not being able to hold on, or more commonly is linked to anxiety and panic. Puppies (and some fully grown dogs) find being left alone incredibly frightening, and can panic and lose control of their bladder or bowel when left. Young puppies should not be left alone, and when you do begin to leave them, this should be something they learn very slowly and at a level they can cope with.

- **The puppy pees when greeting people or when excited.**

  This is again quite common, and again, something they tend to grow out of on their own accord. Keeping greetings (and indeed, life) calm is the best way to help reduce this behaviour, as well as reducing any human behaviour that the puppy may be finding threatening. People in doorways can look quite shadowy and frightening against the light, or may come in with bags, swinging coats, umbrellas, or other things that might seem scary to a puppy.

- **Regression.**

  Your puppy came to you fully toilet-trained, or picked it up very quickly… then suddenly begins to regress. This may be due to too many distractions, or your puppy being too tired, or overly stimulated. Don't panic, just take a step back, keep calm, and put your toilet-training protocol back in place.

If your puppy is suddenly peeing more than normal, and/or drinking more than normal, it is wise to eliminate medical issues with your vet.

SCOUT'S DIARY – DAY 25

'He peed in the park for the first time, which is major!'

Scout in the park. Courtesy Georgia O'Shea.

CASE STUDY: A CAUTIONARY TALE ON CRATE USAGE FOR TOILET TRAINING

Fia, a rescue greyhound who Steph adopted when Fia was three, was a nightmare to toilet train. She had no inclination to take herself outside to wee, and would simply pee on the nearest soft surface – a rug, Steph's bed, or her own bed. Steph had already done a number of dog-training courses at this stage and knew all of the 'right' things to do to toilet train her. Steph brought her outside regularly, rewarded successes, and scatter-fed her in places she tended to relieve herself… but all to no avail.

However, once Steph did some research into life as a racing greyhound, the problem began to make more sense. The dogs will often not leave their kennels from early enough in the evening (perhaps 4 or 5 o'clock), until the next morning. This is a really long time for a dog to hold their bladder, and definitely impossible for a puppy. If this is a regular occurrence, and it is not physically possible for a dog to keep their kennel clean, they may eventually stop trying. So it is most likely that Fia had spent three years without adequate opportunities to toilet outside of her living area and had overridden her instinct to keep her living space clean.

Puppies who are crated can find themselves in a similar position – they cannot move away from their sleeping area to go to the toilet, and if they are getting into the habit of peeing in their beds or 'living area' at a young age, it can be much more difficult to encourage them to be clean in the house.

It took a very long time (about 18 months!) to break Fia of this habit, but with consistent access to the outdoors, and very high value treats for going in the right place, she eventually got there.

JUMPING UP

This is another common puppy behaviour. In the wild, puppies jump up to lick their mother's face to encourage her to regurgitate for them. With us, puppies jump up for attention, to get closer to us, and sometimes to try and take something that we have that they want. If this behaviour is bothering you, or if you're concerned that once the puppy grows up it will no longer be acceptable behaviour, there are some simple steps you can take to discourage it.

Jumping up often happens when the dog is greeting someone. It is sometimes recommended that the dog or puppy should be ignored, but remember that it can be upsetting and frustrating for a puppy or dog if they are ignored when they greet their beloved human parents. It's preferable to greet them calmly and say hello. You may be able to reduce the need for the puppy to jump up by getting down to their level to calmly say hello.

If the puppy has been acknowledged and continues to jump up, you simply turn your back, look away, and use the hand signal (see page 74).

When young puppies jump up at adult dogs, the adult dog will often turn their head away to let the puppy know that they do not particularly welcome this behaviour. Turning your back and looking away if your puppy jumps up at you can have the same effect.

Adult dog Leo looks away as the pug puppy jumps up to lick his face. Courtesy Ali Souza.

Another common concern for puppy owners is that the puppy will jump up at strangers on the street. The difficulty here is that a large proportion of people will be more than happy to greet an adorable puppy that jumps up to say 'hi', and as a result, the behaviour is regularly rewarded. This reinforcement means that, no matter how much training you try to do at this point, the puppy is likely to continue jumping up until the rewarding stops. You can again take away the opportunity for the puppy to practise the behaviour by giving people a wide berth and curving away from them with your puppy as you approach. When Steph was teaching puppy class in London, people often found they didn't have the space to curve sufficiently to prevent this behaviour. In situations where you cannot curve away, she has had some success in breaking the pattern of behaviour by dropping some tasty treats for the puppy to sniff out on the opposite side of you to the one that the passer-by is on. Obviously, this is not feasible on a busy street where you are passing many people, but can be a useful trick to get you out of a bind. However, most of the puppies naturally grew out of this behaviour, because as the dog got bigger, passers-by were less inclined to reward the behaviour, and therefore the dog becomes less likely to practise it.

One thing that has become very apparent to us is that issues, such as the above, which cause puppy parents so much stress and worry, are almost never the issues that people come to us about with adult dogs. Proof that even if it feels like they'll never be toilet-trained, that they'll never stop nipping you and your children, or that they'll be harassing strangers on the street forevermore, these behaviours will be outgrown!

BARKING

Puppies bark rarely, instead they often whine or make other sounds, but as they grow up they start barking in situations where they need to be heard.

All of the sounds they make form part of their language. They are their way of communicating with each other and with us. When dogs use their language, they are trying to say something. They have something on their minds, and we must listen.

Barking is used for a variety of reasons: to express excitement and happiness, or communicate fear and anxiety, frustration and defensiveness. However, they can also learn that barking is an excellent way to get our attention. This can happen if we inadvertently give them attention for this behaviour – even if just to 'shush' them!

It is therefore essential to recognise the context in which the dog is barking. The sound of the barking will be different in the different cases (for example, fear barking is often continuous and punctuated by howling or whining), and may be accompanied by different behaviours.

Some dogs will bark more readily than others, particularly if barking formed part of the job they were bred to do. Some breeds hardly ever bark, and if they do, it tells us that something is wrong in that dog's life.

Barking is often what happens when you didn't respond to the 'small talk' – the calming signals and the body language they may have previously used, and so they feel they have to scream out loud. People do the same – we start shouting and screaming when we feel nobody is listening to us, and for dogs, barking is frequently the same kind of shouting.

Try not to just get irritated when your puppy barks. He has something to say, and it's our job to try and find out what it is. Once we know why they are barking, dealing with the cause will be much more effective than simply trying to stop the barking. For instance, if your puppy is barking whenever you leave the room, it may be a sign that they have separation anxiety. In this case, what you really need to deal with is the separation anxiety, and then the barking will look after itself. If your puppy is barking at other dogs in the park, it could be because they are frightened. Helping them to work on their social skills and to become more comfortable around other dogs will deal with the root of the problem (see pages 128–131), and again, the barking should resolve. Alternatively, your dog might be barking at other dogs because he is used to being able to play with other dogs and is frustrated that he is being prevented from doing so. Providing him with calmer social outlets, and spending time on calm, on-lead walks with other dogs can help alter his expectations and remove the frustration.

You can read more about the different types of barking and how to best manage them in Turid's book *Barking: The Sound of a Language.*

MANAGING LEARNED BARKING

If you find that your puppy is barking because they have learned that it is a great way to get your attention, try looking away and using the hand signal when they start barking, or better still, when you think they're about to start. Wait until at least five seconds have passed since they stopped barking before you give them any attention.

It is important to persevere with this, as often your puppy will have become quite used to the fact that barking gets attention. As a result, when the barking suddenly stops working, your puppy's initial response may be to bark louder, or for longer, or to bark and jump at the same time. It is always tempting for people to presume their method is not working and give up. If you do give up, however, your puppy will have learnt to simply escalate his behaviour if it's not working. So stick with it!

HUMPING

When a puppy is initially brought home and is awaiting the completion of his vaccines, he very often faces an extended period of time without seeing any other dogs. Having been separated from his mum and littermates earlier than nature intended, the puppy is often overwhelmed with excitement when he finally does meet another dog. Often, that is at a puppy class or puppy party, and in these cases you will be sure to see at least one of the puppies getting so excited that he starts humping the others.

Don't worry, this puppy is not a sex pest! It is simply the excitement and stress of the situation causing this behaviour. Likewise, if the puppy is stressed in other situations because there is too much going on, too many things to handle in a new home, too much excitement, frustration, or for any other reason, one of the behaviours you might well see is humping – it may be directed towards people, other animals, beds, cushions, soft toys or other objects. It is a sign of stress, and nothing else.

So, if your puppy starts humping a person, dog, or object, just take him quietly away from the situation. And try to find out what is stressing him, so you can remove that stressor.

Try not to giggle, or pay too much attention to the behaviour, as this can simply encourage the puppy to do it more.

When they reach adolescence, you may see an increase in this behaviour, but again, it tends to happen when the dog has a bit too much going on, and more often than not, they just outgrow this phase so long as we don't pay it too much attention.

MANAGING SEPARATION

Separation distress is a common plight of pet dogs, with as many as 50% suffering from separation anxiety at some point in their lives. Indeed, the RSPCA euthanasia report for 2015/16, showed that 24 puppies were euthanised primarily because they were suffering from separation anxiety.

Dogs suffering from separation anxiety may engage in vocalisations (barking, howling), destructive behaviour, such as digging or chewing, and house soiling in the absence of their human(s). These are not absolute criteria. Your dog can have separation anxiety without these symptoms, but when these symptoms are present, in the absence of the dog's human, the dog almost certainly has separation anxiety.

WHAT MOTIVATES THESE BEHAVIOURS?

Vocalisations: Dogs use howling to call their family members. As outlined, dogs bark in lots of different situations and for different reasons, but the sort of barking found in dogs with separation anxiety is usually fear barking. It is usually quite continuous, sometimes punctuated with whining or howling.

Destructive behaviour: This tends to take two forms – it can be chewing or scraping at exit points, usually doors or windows. This is the dog desperately trying to get out and find you. It can also come in the form of chewing things in the house. Chewing releases serotonin in the dog's brain, so this sort of chewing is the dog's way of seeking some sort of release from their stress.

House soiling: Releasing the bowel or bladder is a stress response linked to the fight-or-flight system. In the wild, releasing this extra weight would allow you to move more quickly!

Indeed, the term separation anxiety is perhaps not the best description of the dog's experience – the degree of 'anxiety' displayed by many dogs suffering from separation anxiety is more akin to panic attacks as suffered by humans. The response can escalate to self-harm in the form of psychogenic licking leading to skin lesions, and dogs can cause themselves injuries in their attempts to escape from the house or a cage to reach their absent human. Although the behaviour that can result from the dog's panic can be rather frustrating for the human too, it is important to remember just how real your dog's panic is and how frightening it is for him. They're not doing any of this to spite you; it's a physiological response to their psychological turmoil.

New puppy parents are often keen to do whatever they can to reduce the chances of their dog suffering from separation anxiety as they get older. However, a lot of the traditional wisdom in this regard is often unhelpful at best, and detrimental at worst. It is commonly suggested that spending too much time with your puppy or spoiling them will lead to separation anxiety. Studies show, however, that 'spoiling behaviours' do not increase the risk of separation anxiety. In fact, several studies provide evidence that dogs not receiving enough attention are actually more likely to display separation issues.

MINIMISING THE RISKS OF YOUR PUPPY DEVELOPING SEPARATION ANXIETY

The most significant thing you can do the reduce the chances of your puppy developing separation anxiety is to help him develop a secure base. As we've already touched upon, dogs do not leave their young to become distressed – to do so would attract predators and alert them to the fact that the puppy was alone. Rather, a care system seems to be in place amongst free-ranging dogs to ensure that the vulnerable puppies are supervised.

Puppies have simply not evolved to be independent enough to be left alone at the age of eight or ten weeks. When you bring home a puppy, it is best if you can plan for the puppy to have round-the-clock care (as you would with a young child) until they are at least four or five months old. First experiences are incredibly formative, and negative experiences even more so, so if you leave your puppy for two hours at ten weeks old, and they get into a state of distress, a pattern of behaviour is already being established, and is more likely to reoccur when they are left alone again.

SCOUT'S DIARY – DAY 22

'Still fretting after me when I leave the room, no matter who else is there. He peed when I left for ten minutes today even though he'd just been to the toilet in the garden...'

Scout anticipates Georgia's departure. Courtesy Georgia O'Shea.

THE UNIVERSAL HAND SIGNAL

We have spoken about the many uses for the hand signal in the 'Good habits' section of this book, so you won't be surprised to hear that you can also use it in your separation training! To use it to prevent separation anxiety in your puppy, start by showing the hand signal before you start moving around the house, to let him know that there is nothing about to happen that he needs to take part in. If you're getting up to go to the toilet, use the hand signal first. If you are going to make yourself a cup of tea, use the hand signal first. If you are going to open or close the window, use the hand signal. Don't worry if your puppy follows you initially. Puppies will follow you around the house, just like small children do, but gradually they become a little more confident about being on their own. The hand signal tells them gently that nothing of interest is happening – and of course you have to be sure that you are consistent about this, and not get a treat for him, or take him out after showing the hand signal. If you have told him nothing is happening, nothing should happen!

MINIMISING STRESS

Increased general levels of stress and anxiety increase the risk of a puppy or older dog suffering from separation anxiety. Encouraging calmness in as many areas of the dog's life as possible and reducing unnecessary stressors can help reduce the chances of your dog suffering from separation anxiety.

Ensuring that your puppy has sufficient quality sleep, plenty of social contact, good food, choices, and appropriate exercise is important in this regard.

A common well-intentioned mistake that people with dogs suffering from separation anxiety make is to try and physically exhaust the dog before they go out, hoping that the dog will be tired enough to sleep when left alone. However, the result of this can be that any fast activity will have left the dog full of adrenaline, and even more primed to react to the additional stressor of being left alone. A nice, calm, sniffy walk, or some calm but mentally stimulating nose work can be much more beneficial, and will leave your puppy tired and calm, rather than tired and wound up!

TRAINING FOR SEPARATIONS

If you have practised the hand signal, it will be easier to teach the puppy to be alone. He knows it means that you are about to leave but will come back, so when you start going out for short periods of time (seconds in the beginning!) and have given him the hand signal first, he will accept it. It has proved to be a great help in helping dogs learn to be alone at home.

Once your puppy is a little older (usually four to five months, but this can vary depending on the puppy), and more independent, you can begin to teach them to be alone for increasing amounts of time. Start leaving them for really short periods of time – seconds rather than minutes. If the dog is coping, you can gradually increase the periods of time you are leaving the dog, but this should always be at a level that the dog can cope with. Always use the hand signal first, don't make a fuss, and move away slowly. When you return, acknowledge the dog, but keep greetings low-key. Having calm departures and returns helps your dog understand that these separations are not a big deal. Completely ignoring them on your return, however, is very upsetting and stressful for them. Imagine how you would feel if your partner came home and totally ignored you when you greeted them?

BEARING IN MIND THE LIMITATIONS...

Dogs are a social species and have not evolved to spend long periods of time in isolation. Although dogs can learn to spend some time alone, it is preferable for most dogs to have some form of company most of the time – human, canine, or in some cases, even another animal will do! Studies have found that dogs in single-dog households are more likely to suffer from separation anxiety.

However, that said, more often than not, acquiring a second dog will not 'fix' the separation anxiety that already exists. But if you are trying to avoid a separation issue, or if you have resolved one, and don't wish for it to recur, it is another step that may improve the dog's overall welfare and well-being.

Another alternative to your dog spending too much time alone is to bring the dog with you! Increasingly, dogs are welcomed into workplaces, where they can increase productivity, improve team cohesion, and reduce stress. Dog-friendly coffee shops, pubs, hotels and restaurants are also becoming more common. A more dog-friendly world reduces the time our dogs have to spend alone. Practising calm sessions (page 71) ensures that your puppy can relax in these places.

> You can read more about taking your dog to the workplace in Steph's book *Office Dogs: The Manual.*

You may have a friend, relative, or neighbour who is at home most of the time and is more than happy to have your puppy as company when you are out. For dogs whose separation anxiety is related to one particular attachment figure, this may not resolve the problem, but again, as with a second dog, it can be a useful tool for preventing a recurrence of resolved separation anxiety, or for preventing it from developing in your puppy.

PAWS FOR THOUGHT: TIME ALONE

In our work, it is not uncommon to meet dogs who are left alone for ten hours during the day and then left to sleep alone overnight for an additional eight hours. This equates to the dog spending three-quarters of their time alone. Think about your routine, and the amount of time your puppy may be expected to spend alone. Are there alternatives? Bearing in mind that dogs are social creatures, can you shift the balance so that they have company most of the time? Can you make a list of suitable people who may be able to keep your puppy/dog company?

TAKE-HOME POINTS FOR CHAPTER 5

- Puppies grow out of most of the most troublesome behaviours of their own accord given time.
- Being tired or overstimulated makes most puppy behaviour worse.
- Puppies don't cope well with being left alone – put puppy-care plans in place, and begin to train for separations slowly once your puppy is four to five months old.
- We recommend avoiding locking your puppy in a crate.

6 *Managing relationships*

COMMUNICATION

Communication is a key to any good relationship, and your relationship with your puppy is no different. If you can learn what your puppy is trying to say to you, and you can respond in a way that your puppy will understand, a harmonious relationship becomes so much more achievable!

Dogs are social animals, so having the ability to communicative effectively between themselves is essential for their survival as a group in the wild. Our pet dogs also retain this system of communication, and it is universal among all breeds, in all corners of the world.

CANINE BODY LANGUAGE

The main methods of communication used by dogs are body language and mimicry, with scent and verbal communication e.g. barking, growling, whining (see page 103 for more information on barking as a form of communication) also playing a significant role. From the very beginning, they will also try to use these communication skills with us, so it is important to be keeping an eye out for them from day one.

Some of the most commonly seen elements of their body language are 'calming signals'. These are so called because dogs will use them to try to calm another dog, person, or situation. They are your dog's way of communicating peaceful intent and diffusing situations they perceive as threatening.

There are many calming signals, but some of the most commonly seen include the following:

- Licking lips.
- Turning or looking away.
- Curving on approach.
- Lifting a paw.
- Yawning.
- Sitting or lying down.
- Sniffing the ground.
- Scratching.

DOI: 10.1201/9781003305156-6

An itch or a calming sign? Courtesy Jolly Doodles.

Here, the poodle looks a bit more worried by the encounter than the Bernese. He shifts his body weight backwards and lifts his paw. Courtesy Stefania Acquesta.

Remember, if your puppy is very stressed or frightened, they will not use their calming signals – just like we can forget about being pleasant and polite when we're stressed!

We see these signals in a wide variety of situations – for example, if we tell our dogs off, or inadvertently do something they find threatening (staring, taking a photograph, patting them on the head), if another dog approaches them, if they feel trapped or confused.

Another thing to bear in mind is that while this language comes naturally to them and they get experience of seeing and using it with their mums and littermates, as with any language, it can become rusty if they do not get to use it for long periods of time. This is another reason why it is so important to ensure that your puppy has plenty of opportunity to spend time with nice adult dogs from the beginning.

Learning to speak! Puppy shows two calming signs: sitting and licking lips. Mum responds by looking away. Courtesy Jolly Doodles.

Adult mentors are a great asset. Courtesy Georgia O'Shea.

More communication! The brindle French bulldog offers a play bow, and the fawn French bulldog responds with a sit and paw lift.

Read more about how dogs communicate in Turid's book, *Calming Signs: On Talking Terms with Dogs.*

HUMAN BODY LANGUAGE

Puppies and dogs pay much more attention to our bodies and movement than to our mouths! In terms of our body language, there are certain 'primate' behaviours that we partake in which can seem very alien, and even threatening to our dogs, such as:

- Bending over them.
- Reaching out towards them.
- Waving arms.
- Hugging.
- Staring at them.

This sort of behaviour will often elicit calming signals from our dogs. So if, for instance, you walk straight at your puppy, bend over him, or pick him up, and you see one or more calming signals, it is important to respond appropriately. Stop what you are doing, or try to find a less frightening way of doing it. If you are bending over your puppy to put his harness on, perhaps you could kneel down, side-on, to do it instead. If you see calming signals when someone else is interacting with your puppy, do help him out.

One of the reasons it is so important to respond to their calming signals is because if these do not appear to be working for them (i.e. if another dog attacks them or if we continue to shout or frighten them despite their calming signals), their communication can become more forceful as they try to make themselves heard: barking, nibbling clothes, biting, biting the lead, growling and more. This communication tends to be clearer to us and we will often respond to it, making the puppy more likely to rely on these less desirable behaviours than on their calming signals.

Of course, as well as inadvertently making our dogs uneasy, we can also use our body language to help our dogs feel more comfortable. Rather than walking straight at them, we can approach in a curve. Rather than leaning over them, we can crouch beside them to stroke them or when we need to groom them. We can help them when they are on-lead by allowing them enough slack to curve, or by curving with them when approaching another dog or person.

As mentioned previously, you can also use your body to communicate important information to your puppy. We mentioned using the hand signal, which is an effective way of letting the puppy know that he is being acknowledged, that you are taking control of things and that he can step down and relax. We also mentioned using turning your back to discourage jumping up, nipping etc.

Simple things like this in daily life will make your puppy feel understood, feel comfortable with you, and your relationship will benefit from it. It takes so little to learn about how they communicate, and to use this knowledge to great effect. If you can open up a two-way dialogue with your puppy, you will reap the benefits – you will feel the difference in being truly together with your dog, being able to communicate effectively and understand each other. The need for shouting and frustration will disappear.

CALMING SIGNALS – RECAP

Calming signals are so called because dogs will use them to try to 'calm' another dog, a person, or situation. They are your dog's way of communicating peaceful intent and diffusing situations they perceive as threatening.

WHEN WILL YOU SEE THEM?

Dogs can use these signals in a wide variety of situations, including:

- Approaching another dog or person to convey friendliness.
- In response to threatening body language such as bending over, reaching out, waving arms, even taking a photo of them.
- If there is conflict between people or dogs, or even if they perceive conflict where there is none – people hugging, for instance, can look like conflict to a dog.
- If someone stares at them.
- If someone tells them off.
- If another dog is being pushy or rude.

What do they look like?

Calming signals include:

- Licking lips.
- Turning or looking away.
- Curving on approach.
- Yawning.
- Sitting or lying down.
- Scratching themselves.
- Sniffing the ground.

HOW TO RESPOND

If they are approaching another dog, just allow them the time and space to use their language. If they are using these signals towards you or another person, look at what they might be finding threatening – is someone in their space? Or approaching head-on? If so, it is advisable to back off a bit. We too can use some of these signals – we can curve as we approach them, avoid staring and towering over them.

PAWS FOR THOUGHT: OBSERVING CALMING SIGNS

Choose one calming signal a week, and look out for it in your dog, in other dogs, in video footage of dogs interacting on the internet. This is a great way of getting your eye in and becoming more fluent in 'dog'!

PUPPIES AND CHILDREN

Many people want their child to experience growing up with a dog, and this is sometimes a motivating factor for people getting a puppy. Sometimes, children and puppies have a beautiful relationship from the very start. In our experience, however, there are often sticking points! Being prepared for these can help ensure things run smoothly, and that both child and puppy have their needs met.

DISPATCHES FROM A DOG TRAINER

Steph had a dog training client with a French bulldog puppy. One of her children's friends picked up the puppy when the mother wasn't in the same room. The child dropped the puppy (or the puppy scrambled free) falling to the ground and breaking his leg. You can imagine how terrible that was for the puppy, and how traumatised everyone was!

This highlights the importance of encouraging respectful interactions between children and puppies.

NIPPING

Children are often upset by puppy nipping, and unfortunately can often be disproportionately subjected to it. This can be for a number of reasons. Commonly, they will engage in more play with the puppy than adults will, and for a puppy, mouthing is an inherent part of play. Children can also react to the nipping in ways that the puppy finds very exciting – they can pull hands and feet away, squeal, run away, hop up and down. All of this is great fun for a probably already over-excited puppy! If your child is getting distressed by the mouthing and nipping, it can be wise to direct them to games that do not involve the same sort of direct, physical interaction with the puppy (see page 56). Teach them that standing still, with their arms across their chest or on their head and looking away (like a 'snooty statue'), is likely to be more effective than flailing about and shouting.

A lot of children will also treat a puppy how they would treat a teddy bear-picking them up, staring into their faces, etc. Most dogs do not particularly like to be picked up, and will often nip to be let down, or pre-emptively nip hands that they think might be about to pick them up. While it is necessary for a puppy to get used to some handling, this is best done in a calm, pleasant way by an adult, rather than by children swooping them up in the air.

JEALOUSY

While many children love the thought of a puppy, we have come across young children who can become jealous of the puppy. There can also be jealousy amongst the children if they feel that the puppy is favouring another child or family member. These may seem like minor issues, but can become problematic in some households. If your main motivation for getting a puppy is your child, consider how much the child will actually enjoy sharing their care-giver(s) with this new family member before making a final decision.

TOYS

As mentioned before, puppies often have a knack for seeking out your most treasured item and having a good old chew on it! The same can go for children's favourite toys, and this can cause great upset. It is wise to make sure any particularly precious toys are kept well out of the puppy's way. As well as causing upset to the child, ingesting the toys or their component parts can pose a risk to the puppy's health.

SLEEP

Most puppies don't get enough sleep, but puppies who live with children almost never do! Houses with children tend to be very active, with lots happening, and the puppy is likely to want to get involved with most of it. Additionally, children will often be tempted to pick up, stroke, or otherwise disturb a sleeping puppy. As a result, the puppies can be overstimulated and difficult to settle much of the time.

To compound problems, all of the above can raise puppy's stress levels and lead to undesirable behaviour and very stressed-out parents!

However, with a bit of preparation you can reduce the strain on everyone and help foster a positive, mutually beneficial relationship between puppy and child.

A calm walk together! Courtesy Shane Ó Cathasaigh.

GET PREPARED

Talk to your children about raising a puppy, and help them learn about what a puppy needs. There are some great resources to teach children about canine body language, and how to interact with dogs.

Of course, it is crucial to your puppy's welfare that any children in his life understand that a puppy is a living, sentient being and that he should be treated with respect and consideration, rather than like a toy. Some small things that children can do to avoid upsetting the puppy include:

- Allowing the puppy to come to them rather than chasing after him or pouncing on him.
- Never disturbing the puppy when he's sleeping or resting.
- Only picking the puppy up with adult supervision. As well as the risk of the puppy wriggling free or being dropped, being held inappropriately can cause damage to your puppy's developing body (i.e. if the legs are held in an unnatural position or the puppy's body is not properly supported).
- Not grabbing things from the puppy. If the puppy has something he shouldn't have, it is better to swap it for something else.
- Not dragging the puppy along on the lead.

SUPERVISION

It has long been known that children and dogs should not be left unsupervised. While it is very unlikely that your puppy could do your child any damage, it is wise to start as you mean to continue – puppies are not long growing up!

What has often been neglected, however, is the fact that there is no point in supervising unless you can see the warning signs that your puppy is getting worried or that things are escalating. It is therefore crucial that whoever is supervising the puppy and child(ren) has a good knowledge of the aforementioned calming signals and can intervene before the situation gets to the point where there is a bite risk.

The vast majority of dog-bite victims are children, and most commonly it is their own dog who bites them. Teaching your child to interact in a respectful way with the dog while he is still young, and to read canine body language, sets them up for a harmonious life together!

ACTIVITIES

Children can be great at creating enrichment toys for dogs. Instead of playing games that will wind the puppy up, redirecting children into making treat parcels in toilet rolls and other recycling is a great way that children can get involved with the puppy in a way that will calm the puppy down.

There are lots of activities children can get involved in with the puppy that won't wind him up. Here, a child hides treats in a brown paper bag for the puppy to find. Courtesy Shane Ó Cathasaigh.

Making a snuffle mat can provide a great toy for your puppy, and is a great task for children to help with. Courtesy Shane Ó Cathasaigh.

Bran tries out the homemade snuffle mat! Courtesy Shane Ó Cathasaigh.

They can also set up treat searches for the dog. They can cut up the treats into really small pieces, and throw them into the grass for the puppy to find. The trick is to keep this calm, and to let the puppy figure it out for themselves. To discourage the children from helping too much, try giving them a stopwatch and notepad so they can time how long the puppy can focus.

Time for a clear out? Let your children take out a selection of old items from the garage or the attic, and lay them out in the garden or a room for the puppy to investigate – just make sure not to include anything that could be a danger to the puppy. You can get more tips on the sort of thing to include in the section on enriched environments.

You can also use art to help your children learn about respecting dogs' space. For instance, younger children can draw a picture of the dog's bed, and mark out the dog's space around this. Nobody can go into this space. Or, they can physically mark this space on the floor using masking tape. If there's more than one child, you can be sure they'll help police each other in this regard!

They can draw up a list of rules for other children coming into the house, with illustrative pictures. This can include rules like not disturbing the puppy when he's sleeping, not picking the puppy up, approaching in a curve, what to do if the puppy jumps up, and so on.

I have often asked an older child to take responsibility for counting how much sleep the puppy is getting to make them aware of the importance of not disturbing their sleep.

However, while children can be involved in really positive ways with rearing puppy, we cannot stress enough how important it is that there is at least one adult who is willing to take full responsibility for the puppy, for the next 12 to15 years (long after the kids may have lost interest or moved out). Children cannot be expected to be responsible for another living being. It's not fair on the child, and it's not fair on the dog.

Spending time together.

CASE STUDY – BRAN THE PORTUGUESE WATER DOG JOINED A YOUNG FAMILY IN DUBLIN IN 2020

Here are some of their thoughts on the experience.

Why did you choose the breed you did?

My wife is Portuguese, and we live here in Ireland, so we wanted to increase the number of Portuguese genes in the house! We also did some research and felt it was a good breed for us, as they don't shed, are supposed to be very family friendly and enjoy some exercise.

A family trip to the beach. Courtesy Shane Ó Cathasaigh.

How did you choose the breeder?

Irish Kennel Club.

How were the early days?

The early days were full of play but bity and jumpy. Although as he was so small it was easily managed. He did cry a bit at night for the first few nights but stopped quite quickly.

What were your favourite things about having a puppy?

His playfulness and his cuddliness.

What were the biggest challenges?

The biting and the jumping.

Is there anything you'd do differently?

In hindsight, Bran is quite a handful given the two kids. We did try and get an adult Portuguese water dog, but couldn't find one. If I were to do it again I would probably hold off till the kids were older or keep searching for an adult dog.

How would you describe things now he's officially an adolescent?

He is calmer than he started off, not as bity or jumpy. But this has been replaced by loud barking and destruction! The hand signal is helping, and even the kids do it!

The universal hand signal is a great communication tool for the whole family to use! Courtesy Shane Ó Cathasaigh.

PAWS FOR THOUGHT – PREPARING FOR DOGS AND CHILDREN COHABITING

Will your puppy be living or regularly interacting with any children? If so, think about your ground rules, and how you're going to ensure that they can get involved with the puppy in a way that will be positive for both the puppy and the child.

MANAGING RELATIONSHIPS WITH OTHER PETS IN THE HOUSEHOLD

People who love dogs often love other animals too, so it is not unusual for people to be introducing their puppy to other resident pets. New puppy owners are often particularly keen that an existing dog or cat will accept the newcomer, as these are the animals with whom they will most likely be sharing their living quarters.

PUPPIES AND ADULT DOGS

This is likely to be one of the easier introductions, and in fact, a sensible adult dog could turn out to be a real asset to you while rearing your puppy. However, at times, puppies can be too much for older dogs, and because they are likely to be living in confined quarters with each other, you will need to ensure that your adult dog has somewhere to escape to when it all gets too much.

OLFACTORY INTRODUCTIONS

When you go to meet your puppy at the breeders, you can be sure that your existing dog will be able to smell what you've been up to! Let them have a good old sniff of whatever you were wearing when you met your puppy. Likewise, you can bring a toy belonging to your existing dog for the puppy to smell.

THE FIRST MEETING

Adult dogs are usually quite accepting of puppies when they meet, so unlike introducing adult dogs, you do not need to worry too much about setting up meetings on neutral territory. You have to be sure the adult dog has met and showed good behaviour around puppies, as some might have issues and it will be too late to find this out when you bring the puppy home. Once you are confident that your adult dog can cope with the puppy, you can simply bring the puppy home, meet just outside the house, and walk into the house together.

PUPPY LICENCE

Adult dogs tend to accept a fair bit of nonsense from puppies under about 16 weeks. They know that they're puppies and don't know any better, so it is extremely unlikely that your adult dog will harm the puppy.

ALLOW THEM TO DO SOME TEACHING

While you do not want your adult dog to frighten or bully the puppy, it is normal for the adult dog to let them know when they're being a nuisance. There is no need to intervene every time the older dog snarls or growls at the puppy – he might just deserve it!

GIVE THE ADULT DOG DOWNTIME

Puppies can really torment older dogs, hanging out of their faces, leaping on them, and chasing them about. Make sure you still spend alone time with your adult dog, and that he has time to rest when and where the puppy can't bother him. If you notice that your puppy is being really relentless with your adult dog and you have to leave them home together, it can be worth using a dog/baby gate to separate the two for at least some of the time. That way they can still see each other and be close if they wish, but if the adult dog needs to get away, he can.

Resident dog Remi lets Scout know that he's overstepped the mark. Courtesy Georgia O'Shea.

HAVE PLENTY OF EVERYTHING

If they're chewing, have more than two chews! If they're playing with toys, have plenty of toys. This can help avoid any resource guarding or squabbling.

PUPPIES AND CATS

Cats are another common pet that puppy owners will need to introduce their puppy to and this multi-species setup is not without its challenges! Puppies will often try to chase cats and cats can be very quick to retaliate with their claws. This can be very stressful for the cat and potentially dangerous for the puppy, who could easily find himself on the receiving end of a claw across the eyes.

How well it works out depends on the human family, the environment, and, of course, the individual animals, but the following practical steps can increase your chances of success.

Keeping a respectful distance and learning to co-exist! Courtesy Georgia O'Shea.

Scent

As with other dogs, you can use scent to allow the resident cat to get a 'preview' of the incoming puppy and vice versa.

Introductions

First impressions count! Choose a time when both animals are nice and calm for your first introduction, make sure both feel they have space to get away if they need to, and consider keeping the puppy on a harness so you can maintain distance if need be.

Making friends? Courtesy Georgia O'Shea.

During Turid's life with animals she has had dogs and cats that lived a very harmonious life together, but also experienced the opposite and all of the extra work and precautions this entailed. If you have a cat-friendly adult dog you could probably take in a kitten, and it will be taken care of by the dog. Once, Turid got a puppy and a kitten at the same time, and they became inseparable friends for life. This also meant that the dog looked at all cats as friends!

On the other hand, if you have a grumpy adult cat, you should definitely reconsider getting a puppy. They can be dangerous to the puppy, and can cause damage with their quick claws!

Barriers

Allowing first meetings through a baby gate or other appropriate barrier takes away the opportunity for any nastiness – the cat can't be pounced upon and the puppy can't be smacked!

Scout bears a war wound on his snout from an encounter with the cat. Courtesy Georgia O'Shea.

Ongoing management

If the puppy is very bouncy around the cat, it is advisable to carry on with more controlled interactions. You may wish to keep the two physically separated by using a baby gate, or you could keep the puppy on a harness and lead around the cat initially so he does not get the opportunity to form a habit of chasing the cat.

What if the two don't accept each other?

If the problems are not too big, and you carefully manage the situation so they do not get the opportunity to bother each other, things may well improve in time, and as the puppy grows up, they may learn to just ignore each other.

But often in these cases it will be necessary to separate them – to divide the house and garden into dog zones and cat zones. If you have a gentle adult cat who is being persecuted by a rambunctious puppy, ensure there are places up high where the cat can go and the dog can't follow.

Our advice is to really consider what is best for the animals, taking into account the personality of your existing cat and their likely reaction to the puppy. We are, after all, the responsible ones and we must try to make the best decision for the whole family, four-legged members included.

PAWS FOR THOUGHT: OTHER ANIMALS

If there's already a resident animal in your house, spend some time thinking about how the new arrival will impact them. How will you manage this? What spaces can be safe spaces for the existing animal to get away from the puppy? Plan some activities that you will do with the resident animal after the puppy arrives so you can make sure they don't feel sidelined. It is good to be prepared, so think about acquiring physical barriers to keep the two apart when you can't supervise, and how you might use these.

TAKE-HOME POINTS FOR CHAPTER 6

- Take time to learn about how dogs communicate and practice reading their body language – this will prevent many misunderstandings!
- Relations between puppies and young children are not always straightforward. It is common for children to take the brunt of the puppy-nipping, and for puppies in a household with children to be over-stimulated and over-tired. Always supervise, encourage calm interactions, and plan in advance how you are going to balance the needs of child and puppy.
- If there are other animals in the household, consider whether they will enjoy the addition of a puppy, and how you can manage the physical space if there is a problem.

7 *Activities to try with your puppy*

Choosing activities to do with your puppy that tap into their natural abilities and instincts will be really rewarding and enjoyable for them. They will not become wound up or over-excited, and most puppies are pleasantly tired and chilled out after taking part in the activities mentioned below.

TREAT SEARCH

When: Soon after a meal when your puppy is not hungry. If your dog is hungry before a game like this, it could be frustrating or stressful for him. You can do this activity daily with your dog.

Where: Preferably somewhere grassy, where no strange dogs are likely to pass by. If you have access to a garden, that's perfect.

What you need: Something tasty and smelly, cut into tiny pieces, no bigger than your baby fingernail. Try cooked chicken.

How: The first time, let your puppy see the treat in your hand and let him see it drop to the ground. He will hopefully pick it up and eat it! The next time, let him see three treats in your hand and let him see them drop to the ground, and the third time five treats.

Once your puppy has this basic understanding, you can throw all your treats into the grass at once and let him sniff around and find them. Do not help him, or praise him, just let him use his nose. You can time how long he can concentrate for and compare it to subsequent attempts.

Why: This is a great activity for your puppy. Using their nose provides great mental stimulation and will tire your puppy out. Sniffing also causes the pulse rate to drop, so it is very calming. And, because it is a task that your puppy is undertaking by himself, and succeeding in, it is good for building his confidence and focus.

DOI: 10.1201/9781003305156-7

A treat search! Ideally, it's best to do this in a secure area without a lead.

ENRICHED ENVIRONMENT

When: When your puppy is small, enriched environments are a great way to introduce him to new things. As enriched environments need regular refreshment, starting off with once a week is ample.

Where: You can do this indoors or outdoors. It should be a safe space where other dogs are not going to get involved.

What you need: Anything that looks, smells, sounds or feels interesting, and that is safe for your puppy to interact with. Do not include things that would frighten your puppy. It's ok if he is a bit wary of a couple of things initially; he can decide to interact or walk away. See Table 7.1 for some of the things you can consider.

How: Take a selection of 10 to 15 items and spread them out so that the puppy can move around them and investigate them. Sit down somewhere silently, and leave your puppy to move around and investigate the items. Don't interfere (unless the puppy is in danger!), or praise or encourage. If he is a bit wary of something, just allow him to decide whether he wants to investigate them or not. If left to his

Table 7.1 Enriched Environment

Smells	Sensations	Sounds	Visual
A cat carrier from a friend	Tarpaulin	Black sacks	A child's pram or another toy on wheels
Another dog's harness or toy	Sand pit	Empty tins	An open umbrella
Used animal bedding from a pet shop	Agility tunnel	Wind chimes	A plastic bag of clothes or scrap material
Hair-cutting from a dog groomer	Hay bales		A stuffed toy
Other people's worn clothes	Tyres		A mannequin
Herbs and spices	Artificial grass		A suitcase
Empty food containers			

own devices he will probably eventually go and check them out. Just leave him be. Again, you can note down how long he spends, what he's particularly interested in, if anything worries him etc. He may come away and take a break, and then return to investigate again. This is fine.

Each time you set up an enriched environment, there should be new things to investigate.

Why: This engages a puppy's natural curiosity and most dogs really like it. Again, it is very stimulating in a calm way. Because your puppy has full choice about what to do, what to investigate, and when to move away, it is good for his confidence. It is a safe way for him to encounter things he will likely come across in life, and will reduce the chances of him being scared of them in real life.

Your puppy will also most likely be exhausted after this, and sleep for quite a while – a bonus for most worn-out puppy parents!

Items that smell of other animals, like this dog's coat, are often of great interest! Courtesy Georgia O'Shea.

An enriched environment including sheep's wool, items of different shapes, a dog bed, dog toys, hen bedding, and more!

Household items that your puppy may not see very often, or even out of context, like this open umbrella and camping bed, can also be included. Georgia O'Shea.

The puppy should be free to interact with the environment in whatever way he sees fit. Courtesy Georgia O'Shea.

Don't worry if your puppy seems a little cautious or worried about any of the items – the important thing is that they're free to move away or investigate at their own pace. Courtesy Georgia O'Shea.

ACTIVITIES WITH OTHER DOGS

Have we mentioned that dogs are social creatures? They need to have the opportunity to have canine friends and interactions with other dogs. Having access to both adult mentors and puppy pals will create a social environment closest to that which they would have in their natural environment.

PARALLEL WALKS

On-lead walks with other dogs are a great way for your puppy to get used to being around other dogs in a calm, sensible way. Here is a basic recipe for organising your own parallel walk for your puppy:

- Start with just one or two, sensible, friendly adult dogs.
- Meet in an open space with the dogs on harnesses and a long enough lead (three to five metres). Give the dogs some time and space to calmly investigate their environment before commencing the walk.
- Start off with a good amount of space between the dogs, and some people as barriers so the dogs are not too excited by the sight of each other.
- Walk along with the dogs parallel to each other.
- As the dogs become more comfortable with each other's presence, you can begin to reduce the distance. If either dog becomes over-excited, increase the space again.

A calm parallel walk with another dog.

Resting calmly together after a short walk.

SEARCH FOR TREATS TOGETHER

This is another calm activity for dogs to do around each other. Spread out so each dog has plenty of space and will not wander into the other dog's treat-search area. Scatter treats for each dog to find.

If either dog is a resource guarder, this may not be appropriate, or you may need to give them heaps of space and barriers in between. You may wish to keep the dogs on a long lead to ensure that they respect the other's space.

Searching for treats together. Courtesy Georgia O'Shea.

VISITING EACH OTHER'S HOUSES/GARDENS

If you have friends or family members with dogs that your puppy has met a few times and gets along with, why not take them to visit each other. Dogs are always really interested to sniff around another house or garden, and hanging out calmly with a canine pal can be an added bonus. It is always best for the dogs to meet out on the street and walk back to the house together, rather than suddenly come nose-to-nose at the front door.

Visiting a friend can be a great adventure. Courtesy Georgia O'Shea.

ACTIVITIES TO THINK TWICE ABOUT...

TRIPS TO THE DOG PARK

Puppies tend to be really delighted to meet other dogs, and people are often tempted to bring them to dog parks to play with other dogs, thinking it will be great for their socialisation.

However, there are a number of potential issues with dog parks:

- Firstly, the play tends to be very fast, and often rough. Your puppy will likely be terribly wound up by this sort of interaction, and chances are that you will have problems settling him later in the day.
- While engaging in these boisterous encounters, your puppy will simultaneously be learning to interact in this way with other dogs they meet in daily life – many of whom will not appreciate this approach. It is much better that your puppy learns how to behave from well-socialised adult dogs in a calm environment.
- There is a high risk that your puppy will have a frightening experience if an already adrenalised dog bounds up to him and tries to play boisterously with him. An early experience like this could have a lasting impact on your puppy's attitude to other dogs.
- There is a risk of physical injury with this 'fast and furious' play.

DOG DAYCARE

In our experience, most dog daycares are not run in a way that is conducive to puppies or dogs being calm and relaxed. The ratio of dogs to humans is often very high, and dogs often spend a lot of time being transported around, before being left largely to their own devices in fields or in a building.

Again, the default can be to leave dogs to play manically and to become very wound up. Walks are often in large groups and can include ball or frisbee chasing (see page 44) and all dogs being yanked along at the same speed. There is often not enough space for dogs who do not want to partake in this sort of activity to have a quieter time.

In a well-run daycare, there will be clear structures in place, with plenty of allocated time for resting. The normal dog behaviours that we have looked at in this book should be considered – dogs should be able to sleep socially (with human and/or canine company depending on their preference), with plenty of choice of comfortable beds. For puppies, these sleeping areas should also be warm enough for them. A large proportion of their waking time should be spent in calmer activities such as nice, on-lead social walks, and exploring interesting environments. The physical environment will be such that the dogs have things to get involved in other than constant playing, and plenty of space to be apart from each other if they so choose. Play should be in small, politely introduced and well-matched groups and not be allowed to escalate into a frenzy. If these groups can be stable, with the same few dogs together each day, all the better. Puppies and seniors, who will have different needs to young, healthy, fit dogs, should have a daily routine that reflects these needs.

If you cannot find a really top-notch daycare, it is much better for your dog's social requirements to arrange a babysitter, and to meet with dog friends for some of the aforementioned activities.

The only time that a less-than-ideal dog daycare might be the lesser of two evils is if your dog is suffering from separation anxiety, no alternative care is available, and your dog will find it more distressing to be alone than to be in daycare.

TAKE-HOME POINTS FOR CHAPTER 7

- Providing adequate mental stimulation to your puppy is key to him developing into a happy, confident dog.
- It will also tire him out in a nice, calm way.
- Choose activities that tap into his natural instincts – to sniff, to investigate, and to forage.
- Opting for calm, well-managed activities with other dogs and puppies will help ensure your puppy has the opportunity to develop good social skills that will stand to him for life!

8 *Beyond puppyhood – what to expect next*

From about five to six months, your puppy ceases to be a puppy, and becomes a 'young dog'. From here until about a year, they can become a little unfocused! Then from about 12 to 18 months, they are in the dog equivalent of the 'terrible teens'! In some ways, this can be an even more challenging period than puppyhood – it is no coincidence that the most common age for dogs to be surrendered to rescue kennels is during this adolescent phase! In fact, a recent study into adolescent dogs showed that they display similar behaviours to adolescent humans.

So, we didn't want to leave you without a few words of warning about what to expect, and some tips for getting through this stage. If you've laid the foundations of a good relationship during the puppy phase, however, you'll be heading into this phase on the best possible foot!

SECOND FEAR PERIOD

Dogs will often experience a second fear period during adolescence. From an evolutionary perspective, this makes sense – as a puppy is getting older and more independent, a degree of extra caution can prevent them from ending up in dangerous situations when mum is no longer around to look out for them.

You might find that your adolescent dog is suddenly startled by things that never worried them before – a bag blowing down the street, children, the hoover.

As with the first fear period, don't force anything. Allow your dog the opportunity to choose whether or not to approach anything that frightens them. If they choose to move away, don't worry. Give them the choice again another time, and they will probably eventually choose to investigate.

If you become aware that your dog might be experiencing a fear period, try not to introduce too many new things during this time, and keep things calm and quiet. Like the first fear period, this will pass in time. A little support and patience is all that is required!

DOI: 10.1201/9781003305156-8

SCOUT'S DIARY – SEVEN MONTHS OLD

'Scout turned seven months, and overnight developed a fear of... bushes! Although he had walked in the same landscape multiple times without incident, he suddenly started avoiding standalone bushes and trees. He would freeze, bark, lunge, and become very agitated. He would appear to be very shaken afterwards and would require quite a lot of time to decompress.

These reactions weren't consistent though – often a walk would go by and he wouldn't even notice the standalone bushes and trees.

These incidents continued for about a month, and then ceased as quickly as they had begun!'

Teenage Scout, having recovered from his transient fear of trees! Courtesy Georgia O'Shea.

INCREASING INDEPENDENCE

One day, you had a puppy who followed you around devotedly, kept an eye on your every movement, and never left your side. The next, your puppy runs away from you in the park and seeks out the company of strangers, other dogs, or whatever else takes their fancy! Just like with human teenagers, the changes taking place in your dog's brain make them seek out independence. They are less likely to look to you for guidance, and more likely to make their own decisions.

This desire to move away and investigate is a normal part of their development, and is designed to prepare them for adult life – accepting that this is a normal part of development is important. Try not to get cross with your dog – provide them with appropriate opportunities to make choices and to investigate in safe environments. This will be good for their confidence and their sense of well-being. If we continue to manage them as puppies, they are likely to become frustrated – just as a human teenager would if you wouldn't allow them to stand on their own two feet a bit more!

You can put management in place to prevent your young dog coming to harm – for instance, if they have taken to running off you may wish to keep them on a long lead in unenclosed spaces. If they are being rambunctious and over the top with other dogs, you may wish to keep them away from dog parks, lest they develop bad habits! A well-run social walk can provide a better environment for social interactions.

PUSHING BOUNDARIES

It is also normal for young dogs in adolescence to become a little deaf to what their human parents say. During the adolescent period, they may ignore requests that they would have gladly complied with during their puppyhood. They may also seem difficult to train. Again, refrain from becoming angry with your dog – this is all part of growing up, and a passing phase.

Interestingly, it has been found that dogs with a less secure attachment are more likely to exhibit these behaviours, so if you put in the leg-work in creating a secure attachment early on in your relationship with your puppy (being present and not leaving them alone to become distressed, avoiding punishment, meeting their physical and emotional needs), you might find this is less of an issue.

RISK-TAKING

Your dog is at an age where he might naturally be moving away from his family group, an inherently risky business! Risk-taking behaviours become more apparent in adolescence and may even manifest as the dog showing more 'aggressive' tendencies. These tendencies, when they manifest in adolescence, tend to be transient and resolve of their own accord. The second fear period can also be responsible for a dog appearing be more defensive around other dogs, people, etc.

SEXUALLY DIMORPHIC BEHAVIOURS

You will probably notice that your male dog will start lifting his leg when he urinates, when he previously squatted. They do this to keep themselves clean, and at this age, they have the balance and strength to do it. He may engage in urine marking (peeing just to leave a message rather than because he needs to go – you might notice him peeing on every lamppost, and on people's chairs when he goes visiting!) or may become more likely to roam in search of female dogs in heat.

Female dogs will probably come into season between 6 and 12 months. When this happens, she may experience some bleeding, she may urinate more than usual, or seem more eager to go out. If you suspect that your dog is in heat, exercise caution and keep her away from any male dogs.

FUN FACT!

Have you ever seen a female dog cock her leg?

One theory about this is that if a female puppy is carried in the uterus in-between two male puppies, she may end up being exposed to an increased level of testosterone, which can lead to her displaying some typically 'male' behaviour such as leg-cocking and scent-marking.

BRAIN DEVELOPMENT

Remember, your puppy's brain will not be fully developed until they are quite a bit older. In humans, the development of the prefrontal cortex (the part of the brain responsible for higher brain function such as decision-making) continues into our mid-twenties. Similarly, the maturation of a dog's prefrontal cortex carries on well into the adolescent period. Therefore, even when a dog appears fully grown, they may not be fully mentally mature. As a result, young dogs, (like young humans!) can be reckless, and do not always make the best decisions – so do cut them some slack!

NEUTERING

Neutering is a contentious topic, but it is one that we are often asked about when working with puppies. If you have rescued a puppy, it may be part of your agreement with the rescue that you will neuter your dog at six months. For those who buy a puppy, your vet may start encouraging you to neuter your puppy at around this age too. However, there are also countries, such as Norway and Sweden, where you are only allowed to neuter your dog if there are clear medical indications for doing so.

This is largely a medical issue, and it is good to see that increasingly, vets seem to be encouraging their clients to adopt a 'wait and see' approach, and making decisions on a case-by-case basis, taking the puppy's personality into consideration. If your vet is encouraging you to neuter at six months simply because they encourage everyone to do this, do give it some thought and weigh up the pros and cons. After all, there's no going back once you neuter!

The most obvious benefit of neutering is that it effectively eliminates the risk of your dog bringing unwanted puppies into a world which already has a saddening number of such dogs. It also reduces the risk of some diseases, whilst also increasing the risk of others (Table 8.1). There is a lot of research being done currently into the impact of neutering on dogs, and some of this research is included in the further reading section. Obviously, you can discuss these further with your vet, however it may worthwhile doing some of your own reading beforehand.

Table 8.1 Risks of Neutering

Female dogs' risks reduced	Female dogs' risks increased	Male dogs' risks reduced	Male dogs' risks increased
Mammary tumours (especially if neutered before two and a half years)	Osteosarcoma (risk doubled when neutered before one year)	Testicular tumours or torsions	Osteosarcoma (risk doubled if neutered before one year)
Pyometra (potentially fatal uterine infection)	Splenic, cardiac, and mast cell tumours (risk multiplied by five)	Peri-anal adenomas and fistulas	Splenic, cardiac, and mast-cell tumours
Peri-anal fistulas	Urinary tract tumours (risk doubled)	Groin hernias	Prostate and urinary tract tumours
Uterine, cervical, and ovarian tumours	Urinary tract infections and incontinence	Benign prostatic hypertrophy	Age-related cognitive dysfunction (canine Alzheimers)
	Obesity, orthopaedic problems (such as hip hysplasia and osteoarthritis)		Orthopaedic problems (such as hip dysplasia and osteoarthritis)
	Hypothyroidism		Immune-mediated disorders (increase less marked than in female dogs however)
	Immune-mediated disorders		

If you wish to read more about the impact of neutering on health, a review of the various studies looking at the impact of neutering on health was conducted by Anita M. Oberbauer, Janelle M. Belanger, and Thomas R. Famula (2019), and this is a good place to start your research in this area.

From a behavioural perspective, there are a number of things we would ask you to keep in mind when making this decision.

- The sex hormones affected by neutering are responsible for much more than just reproduction. They also play a significant role in modulating the immune system, in bone health, and in brain development.
- When, and not just whether, you neuter is significant. Differences are observed between an early neutering strategy versus a delayed neutering strategy (i.e. later than two years old).
- There is little evidence that neutering positively impacts a lot of the behaviour it is reputed to ameliorate, such as excitability, reactivity, or aggression.
- Neutering may be effective in reducing behaviours such as urine marking (this is separate to toilet training), roaming in search of female dogs, mounting, and fighting.
- The loss of testosterone, which is inevitable when male dogs are neutered, can lead to a dip in confidence and the development or increase in fear-based behaviour. Studies suggest that neutering between 7 and 12 months markedly increases the risk of aggression directed towards strangers.
- Neutering may also increase the risk of your dog developing separation anxiety.

DISPATCHES FROM A DOG TRAINER

The most common behavioural fall-out we have seen in neutered male puppies is an increase in fear-based behaviour, such as barking at strangers, chasing joggers, etc.

So, while it is a complicated subject, and each dog and situation is different, our personal advice is for people to consider not rushing into neutering a puppy. From a behavioural perspective, there is little to be gained, and nervous male dogs are the ones we would expect to have the most negative outcome from neutering. If you do decide to neuter your dog, try to choose the optimum time in terms of their physical and emotional development (and in the case of female dogs, their oestrus cycle). If you decide not to neuter, or to delay neutering, remember that you must be certain that you can practice responsible dog ownership, so as to avoid any unwanted litters!

TAKE-HOME POINTS FOR CHAPTER 8

- As with human youngsters, adolescence with your puppy can be challenging!
- Expect to see a second fear period during which your puppy may react fearfully to new things, or indeed, to things that never bothered him before.
- Your puppy may strive for independence and be less interested in sticking with you while out and about. Manage these situations with a long lead and careful walk planning!
- Physical changes may occur and you might notice your male puppy cocking his leg when he urinates, or notice your female dog go into heat.
- The canine world is becoming increasingly aware that neutering is a complex issue that can impact both behaviour and health, while at the same time being an effective method of preventing unwanted litters. Spend some time researching this topic before deciding when and if to neuter your puppy.

9 *Turid's case studies*

There is no such thing as a perfect puppy, or as being prepared for all the problems that can pop up during a dog's life. Many things have an effect on a living creature – their environment, their health, their experiences, and, of course, their individual personality. At some point, everybody is taken by surprise by their puppy or dog, or suddenly meets a situation they were not prepared for, and how it is handled in the moment can result in something unexpected. We all will sometimes choose the least clever way to handle these situations (and sometimes the best way!). The mistakes we make can always be mended; the important thing is to admit that we should maybe have made another choice, and then turn it around. When people have called me about a problem they have, they have often waited so long that it has become a deep and almost impossible problem to solve. At best, it will require much more time, work, and planning to get around.

My advice is that if you begin to think you might have a little problem, you have. Or your dog has! Just be open about it and do something before it gets difficult. The faster you react, the faster it will be solved, with minimal effort and cost.

Throughout my 50 years as a dog trainer, I have had thousands of puppy cases, and have therefore come across almost all of the problems imaginable. I have picked out a few to illustrate some of the things that can happen, and what to do about it.

COCKER SPANIEL, FOUR MONTHS OLD

Problem: When meeting people, bicycles, dogs – actually everything – he would start to walk slowly, freeze in a sit or lie-down position, and start barking.

Puppies are curious, and want to check things out, so this reaction made me suspect that he was overwhelmed. After a little detective work it turned out that his walks had been too long for such a small puppy, and the humans had been walking too fast. He had had to run to keep up and wasn't given the chance to explore anything, which is vital for puppies.

It had all been too much for a puppy to handle.

Solution: By starting with much shorter and slower walks in quieter places, and letting the puppy look around at things from a bit of a distance, to explore and take his time, it didn't take long before he had calmed down and got back his natural puppy curiosity.

At this point, it is crucial to keep going slowly, and not to start rushing again. If the puppy stops, walks slowly, or gets into a freeze position, just let him do it. No

DOI: 10.1201/9781003305156-9

attention, no luring (showing the puppy a treat and getting him to follow it), no talking. Just wait. If a puppy reacts to something, let him choose what to do; that is the absolutely best way of learning to cope – and learning to cope for life.

LABRADOR RETRIEVER, FIVE MONTHS OLD

Problem: The puppy cannot be alone at home without chewing the sofa, books, and pillows.

Of course not. Very occasionally, a puppy can learn quite early to accept being alone at home, but for most of them, they are not able to at this age. It takes a lot of confidence for puppies to be alone – just as it does for young children. You would not leave a toddler alone, and a puppy should not be left alone either. They cannot take responsibility for themselves. They need an adult to do that.

Solution: The best thing to do is to get the puppy a dog-sitter, and after a short period of calming down and getting their confidence back (perhaps a couple of weeks), slowly start the preparations for learning to be alone at home again.

The dog-sitter is not forever, but it will depend on the puppy how long you will need one for.

It is also important not to do stressful things with the puppy in the meantime (fast play, lots of excitement, frightening experiences), as this will just create problems for him learning to relax. The walk just before you leave should be calm and short.

In this case it took about two months before she was comfortable being alone at home. Sometimes it will take longer.

DACHSHUND, FOUR MONTHS OLD

Problem: The owner had just become a grandmother for the first time, and the puppy got hysterical when the baby was present. First, I had to find out what exactly that meant. What did the puppy do, and when? We talked at length about this and I got the picture. The puppy got frightened when the owner picked up the baby and hugged and kissed him, and she also scolded the puppy when he started to bark – making matters worse!

Dogs, no matter what age, get nervous when their family members meet strangers and get too close, hand-shaking, hugging, dancing, and sitting too close. As this is not normal dog behaviour, they tend to misinterpret it as conflict, and feel that it must be stopped. They will try to come in between the people to split them up, or try to make them move apart in some way. In this case, the puppy did not manage to create space between his owner and the baby, being very small-sized, and also young, (too young to take any responsibility). So, he became hysterical because he thought the situation was dangerous.

Solution: For a short time the baby did not visit his grandmother, while she practised the hand signal (see page 74) in other situations, and got the puppy a little more used to baby noises by walking for short periods of time with the baby and his mum when they were out in the pram. He also heard the baby in other parts of the house. The owner paid no attention to these noises, and just used the hand signal when the dog seemed to react a bit. They had no access to other children's noises, so in this case that was not an option. After a couple of weeks, the grandmother, baby, and puppy started to spend time in the same room briefly, but with no hugging and carrying the baby around yet. That came soon after. Giving the hand signal, taking up the baby for a short amount of time, and increasing this gradually, and things started to settle. The puppy was also allowed to investigate the baby, and was invited to be present when the baby was being fed.

Dogs usually understand, when a baby arrives, that they belong to the family. The family smell will tell them, and they accept it naturally. They understand that babies must be looked after. In this case the puppy did not understand why he was told off and could not be with the baby, and everybody else could. He did not understand, and got scared. Turning things around a bit and including him in the family group did the trick.

GERMAN SHEPHERD, FIVE MONTHS OLD

Problem: The humans took the puppy to groups of dogs playing in the park, and he did not show or respect any calming signals (see page 103). At home, they said, he used the signals, and seemed to understand them, but as soon as he came out and started to play, they were forgotten. It always ended with him chasing one of the small dogs and scaring them.

This is exactly how bullies are created, so that had to stop. Playing arouses dogs. The stress gets higher and higher, and in a group of puppies, if an adult dog is present, she would stop it within seconds. That is why in puppy classes I always had one of my adult dogs with me, and she would efficiently stop the playing before it reached too high a level of excitement. If we have no adult dog to do it, we have to do it ourselves. Aroused playing creates bullies and dogs being afraid of other dogs, and is not necessary.

Here the problem was already established, but the strategy was simple: meeting nice, well-socialised adult dogs who would teach him what to do; also having brief meetings with one other puppy of his own size and age and letting them take a short walk together, rather than standing around letting them play. During a walk there should be other things to take an interest in too, other than just playing.

As he grew up, he was able to meet with other dogs of different ages and sizes, but the focus was always on calm walking together. If he was playing, and things began to get a little too much, the play was interrupted before bad habits could be re-established. In time, he turned out to be a nice, social gentleman.

RHODESIAN RIDGEBACK, THREE MONTHS OLD

Problem: The puppy had started to bite the leash when they walked him, and, more recently, had also begun biting the owner's trouser legs.

This is the most common way for a puppy to tell you that something is too much. He had probably tried to tell them in other ways that he was tired and could not walk anymore, but since nobody recognised this, it led to biting.

In this case the walks were far too long, and the puppy's muscles just weren't strong enough to do it. They had also walked too quickly for his short legs, and he had to run the whole time to keep up. It takes time to build muscles, and no puppy has enough strength at that age for constant running on their walks.

Solution: At that age we recommend not 'walking' dogs at all. We recommend taking them out and letting them move freely around so they can build balance and muscle gradually. When it comes to on-lead walking, start walking them for just a few minutes at the start, in places where there are variations of terrain and much to explore. Walk slowly enough for them so they do not need to run and get exhausted. Then build the length of the walks as they grow up, but keep the pace slow enough for them to walk (not run!) and be curious, to sniff, and to explore the environment.

As soon as the owners understood this, the biting stopped, of course.

IRISH TERRIER, THREE AND A HALF MONTHS OLD

Problem: The owner was really worried about the puppy not eating much, and being very thin.

In this case it was obviously very difficult for the puppy to chew the kibble they gave him. Puppies have soft gums, and small teeth – they just cannot chew the hard kibble without getting sore mouths. As we've mentioned, mother dogs regurgitate food for young puppies, and you sometimes even see unrelated adult dogs automatically regurgitating food as soon as they see a puppy of that age.

Solution: We must learn from dog mums; they know best when it comes to rearing puppies. As we've mentioned, puppies need softer food till they start to get real teeth to chew with, and even then they should not need to chew hard food all the time – they just cannot eat it without problems.

When this puppy got food made for him that was much softer, he started to eat, enjoy his food, and put on weight properly.

10 *Closing remarks*

As we reach the end of our journey together, there are a few thoughts with which we would like to leave you.

Raising a puppy is no easy feat – it's tiring and stressful at times, and many people doubt themselves (and their puppy!) in the process! But nothing worth doing is ever easy, and committing to the process and all of the challenges it presents will pay dividends.

Remember first and foremost that your puppy is a sentient, thinking, capable being. Like us, he has just one life, and part of our job is enabling him to make the most of that – to have as great a degree of autonomy as he can safely have, to have his physical and psychological needs met (even when they're at odds with our own desires!), and, simply put, to be a dog.

Of course, you can, and should, offer guidance, and provide him with the skills he needs to lead a fulfilling life – a dog who is calm and confident enough to behave appropriately and has developed good habits will have many more opportunities than the dog who has not.

We hope that you will keep learning about dogs, and more specifically, about your dog – that you will invest time getting to know him for who he is, learning how he communicates with you, what he likes and dislikes. We hope you will try to enrich his life with the former and help him to avoid the latter!

Most importantly, we hope that you will value companionship over compliance and that you and your puppy will live a happy, harmonious life together, as partners and as friends!

DOI: 10.1201/9781003305156-10

An update from Scout – now eight months old

SLEEPING

Scout has become a great fan of his bed (and any other bed he comes in contact with). He will often have a lie-in after I get up in the mornings – to make sure he is fully charged for the day ahead! He is comfortable sleeping in lots of different places, with other people and dogs around… and even in canoes!

EATING

Scout has an ever-growing appetite! He is still on a kibble-based diet and this is being supplemented by 'toppers' of homemade meals giving his diet lots of extra textures and flavours. We plan to transition him to a raw or cooked-meat diet soon.

He has ostrich bones, beef trachea, and turkey necks as chews although his teething seems to have subsided anyway.

EXERCISE

Scout isn't in a strict '2 walks a day around the park' routine; we have instead opted to go for the 'he goes where we go' tactic. Often this can mean Scout doesn't get out till after lunchtime at the weekends but has always had some activity dedicated to him during the time at home. We have found that he can come home exhausted from just watching us work on the boat or mountain bike (we take turns minding him at the café).

He has been given some time to run and explore, using a long line, as his recall and my confidence in managing his interactions are still being worked on and we are still endeavouring to protect his confidence around 'pushier' dogs. We allow Scout to exercise at his own pace and explore lots of new smells and surfaces along the way!

Scout is still attending weekly swimming lessons at a local hydrotherapy centre. He adores his time at the pool and we are so glad we gave him the opportunity to attend… it has created a very confident water dog who has been yacht sailing, canoeing, and surfing!

ALONE-TIME

Scout has had a couple of opportunities to spend time alone at home, albeit with Remi. We also purchased a baby monitor to help with our human anxiety! He tends to bark and whine for a few minutes and quickly falls asleep… It is something that we need to make an effort with, for his sake, so that he is comfortable being alone for short periods.

SOCIALISATION AND HABITUATION

Scout continues to have carefully selected social interaction with other dogs and environments. Scout has met lots of older dogs who we hope will show him appropriate play and behaviour. He has been on road trips, built relationships with different people, been around children and other animals, slept, eaten, and played in new places… and even managed to listen and engage with me when there was a big world to explore.

This element of Scout's training has been somewhat curtailed by the ongoing pandemic.

FEAR PERIODS AND PUBERTY

The week Scout turned seven months we saw some signs of puberty appear. He began to regress in some areas of training. He began mounting inanimate objects and lifting his leg when he urinates. Luckily, aside from this, he has maintained his sweet personality. He has experienced a fear period recently and he has barked at the occasional person (usually only when he is joining in with Remi) so we hope to continue to build his confidence and get through these times.

IN SUMMARY

Scout is in many ways a typical puppy and we are working hard to be consistent in our expectations. We are giving him new experiences whilst maintaining some pillars of routine or parts of his 'default day' so he is an adaptable little guy. We feel he is mature for his age, communicates well with other dogs, and is enjoyable to be around… although he has his moments of madness! We continue to be aware of the bigger picture and set him up for success where possible by giving him coping skills. We advocate for him with other dogs whilst letting him take calculated risks… but no matter what, he is still our best boy.

References and further reading

BY THE AUTHORS

Rugaas, T. (2007). *Barking: The Sound of a Language*. Dogwise Publishing.
Rugaas, T. (2006). *Calming Signs: On Talking Terms with Dogs*. Dogwise Publishing.
Rugaas, T. (2004). *My Dog Pulls, What Should I Do?* Dogwise Publishing.
Rousseau, S. (2019). *Office Dogs: The Manual*. Hubble & Hattie.

FURTHER READING

Boitani, L., Francisci, F., Ciucci, P., & Andreoli, G. (1995). Population biology and ecology of feral dogs in central Italy. In J. Serpell (Ed.), *The Domestic Dog: Its Evolution, Behaviour and Interactions with People*. Cambridge University Press, 217–244.
Bradshaw, J. (2012). *In Defence of Dogs*. Penguin.
Brady, C. (2020). *Feeding Dogs: The Science Behind the Dry Versus Raw Debate*. Farrow Road Publishing.
Coppinger, R., & Coppinger, L. (2002). *Dogs: A New Understanding of Canine Origin, Behavior and Evolution*. University of Chicago Press.
Eaton, B. (2011). *Dominance in Dogs: Fact or Fiction*. Dogwise Publishing.
Kvam, A.-L. (2011). *The Canine Kingdom of Scent*. Dogwise Publishing.
O'Heare, J. (2005). *Dominance Theory in Dogs*. DogPsych Publishing.
Pal, S.K., Ghosh, B., & Roy, S. (1998). Dispersal behaviour of free-ranging dogs (Canis familiaris) in relation to age, sex, season and dispersal distance. *Applied Animal Behaviour Science*, 61(2).
Paul, M., & Bhadra, A. (2016). Do dogs live in joint families? Understanding allo-parental care in free-ranging dogs. *PloS ONE*, 13(5), e0197328.
Robertson, J. (2022). *How to Build a Puppy Into a Healthy Adult Dog*. CRC Press.
Robertson, J., & Mead, A. (2013). *Physical Therapy and Massage for Dogs*. CRC Press.

REFERENCES

ADOLESCENCE

Asher, L., England, G., Sommerville, R., & Harvey, N.D. (2020). Teenage dogs? Evidence for adolescent-phase conflict behaviour and an association between attachment to humans and pubertal timing in the domestic dog. *Biology Letters*, 16(5), 20200097. https://doi.org/10.1098/rsbl.2020.0097

ATTACHMENT

Mariti, C., Lenzini, L., Carlone, B., Zilocchi, M., Ogi, A., & Gazzano, A. (2020). Does attachment to man already exist in 2 months old normally raised dog puppies? A pilot study. *DOG BEHAVIOR*, 6(1), 1–11.

CHOICES

Foa, E.B., Zinbarg, R., & Rothbaum, B.O. (1992). Uncontrollability and unpredictability in post-traumatic stress disorder: An animal model. *Psychological Bulletin*, 112, 218–238.

Lickerman, A. The desire for autonomy: Why our need to make choices freely is so crucial to our happiness. https://www.psychologytoday.com/ie/blog/happiness-in-world/201205/the-desire-autonomy

McMillan, F.D. (2019). The mental health and well-being benefits of personal control in animals. In F.D McMillan (ed) *Mental Health and Well-being in Animals*. CABI Publishing.

Rock, D. A sense of autonomy is a primary reward or threat for the brain: Why employees (and your kids) sometimes lose the plot. https://www.psychologytoday.com/ie/blog/your-brain-work/200911/sense-autonomy-is-primary-reward-or-threat-the-brain

Rodin, J., & Langer, E.J. (1977). Long-term effects of a control-relevant intervention with the institutionalized aged. *Journal of Personality and Social Psychology*, 35(12), 897.

CRATES

Lindsay, S. (2000). *Handbook of Applied Dog Behavior and Training: Adaptation and Learning*, Vol. 1. John Wiley & Sons.

EARLY WEANING

Moriah Hurt, B.S., Stella, J., & Croney, C. (2015). Implications of weaning age for dog welfare. https://extension.purdue.edu/extmedia/VA/VA-11-W.pdf.

E-COLLARS

China, L., Mills, D.S., & Cooper, J.J. (2020). Efficacy of dog training with and without remote electronic collars vs. a focus on positive reinforcement. *Frontiers in Veterinary Science*, 508.

FOOD

Bradshaw, J.W. (2006). The evolutionary basis for the feeding behavior of domestic dogs (Canis familiaris) and cats (Felis catus). *Journal of Nutrition*, 136(7), 1927S–1931S.

Butler, J.R.A., & Toit, J.T. (2002). Diet of free-ranging domestic dogs (Canisfamiliaris) in rural Zimbabwe: implications for wild scavengers on the periphery of wildlife reserves. *Animal Conservation*, 5(1), 29–37.

Korda, P. (1972). Epimeletic vomiting in female dogs during the rearing process of their puppies. *Acta Neurobiologiae Experimentalis*, 32(3), 733–747.

Lippert, G. & Sapy, B. (2003). *Relation between the Domestic Dogs' Well-Being and Life Expectancy.*

LONGER LEADS

www.dogfieldstudy.com

NEUTERING

Anderson, K.L., Zulch, H., O'Neill, D.G., Meeson, R.L., & Collins, L.M. (2020). Risk factors for canine osteoarthritis and its predisposing arthopathies: a stematic review. *Frontiers in Veterinary Science*, 7, 220. https://doi.org/10.3389/fvets.2020.00220

McGreevy, P.D., Wilson, B., Starling, M.J., & Serpell, J.A. (2018). Behavioural risks in male dogs with minimal lifetime exposure to gonadal hormones may complicate population-control benefits of desexing. *PLOS ONE*, 13(5), e0196284. https://doi.org/10.1371/journal.pone.0196284

Oberbauer, A.M., Belanger, J.M., & Famula, T.R. (2019). A review of the impact of neuter status on expression of inherited conditions in dogs. *Frontiers in Veterinary Science*, 397.

Starling, M., Fawcett, A., Wilson, B., Serpell, J., & McGreevy, P. (2019). Behavioural risks in female dogs with minimal lifetime exposure to gonadal hormones. *PLOS ONE*, 14(12), e0223709. https://doi.org/10.1371/journal.pone.0223709. eCollection 2019.

PUNISHMENT VERSUS REWARD-BASED TRAINING

Blackwell, E.J., Twells, C., Seawright, A., & Casey, R.A. (2008). The relationship between training methods and the occurrence of behavior problems, as reported by owners, in a population of domestic dogs. *Journal of Veterinary Behavior*, 3(5), 207–217. https://doi.org/10.1016/j.jveb.2007.10.008

China, L., Mills, D.S., & Cooper, J.J. (2020). Efficacy of dog training with and without remote electronic collars vs. a focus on positive reinforcement. *Frontiers in Veterinary Science*, 7, 508. https://doi.org/10.3389/fvets.2020.00508

Herron et al. (2009). Survey of the use and outcome of confrontational and non-confrontational training methods in client-owned dogs showing undesired behaviors. *Applied Animal Behaviour Science*, 117(1–2), 47. https://doi.org/10.1016/j.applanim.2008.12.011

Hsu, Y., & Sun, L. (2010). Factors associated with aggressive responses in pet dogs. *Applied Animal Behaviour Science*, 123(3–4), 108–123. https://doi.org/10.1016/j.applanim.2010.01.013

PUPPY FARMS

McMillan, F.D. (2017). Behavioural and psychological outcomes for dogs sold as puppies through pet stores and/or born in commercial breeding establishments: Current knowledge and putative causes. *Journal of Veterinary Behavior*, 19, 14–26.

PUPPYHOOD AND CANINE PARENTING IN NATURE

Pal, S.K. (2005). Parental care in free-ranging dogs, Canis familiaris. *Applied Animal Behaviour Science*, 90(1), 31–47.

Paul, M., & Bhadra, A. (2017). Selfish pups: Weaning conflict and milk theft in free-ranging dogs. *PLOS ONE*, 12(2), e0170590. https://doi.org/10.1371/journal.pone.0170590

Paul, M., & Bhadra, A. (2018). The great Indian joint families of free-ranging dogs. *PLOS ONE*, 13(5), e0197328. https://doi.org/10.1371/journal.pone.0197328

Pongrácz, P., & Sztruhala, S.S. (2019). Forgotten, but not lost-alloparental behavior and pup-adult interactions in companion dogs. *Animals: An Open Access Journal from MDPI*, 9(12), 1011. https://doi.org/10.3390/ani9121011

Solomon, J., Beetz, A., Schöberl, I., Gee, N., & Kotrschal, K. (2019). Attachment security in companion dogs: Adaptation of Ainsworth's strange situation and classification procedures to dogs and their human caregivers. *Attachment and Human Development*, 21, 389–417.

SENSE OF SMELL

Quignon, P., Kirkness, E., Cadieu, E., Touleimat, N., Guyon, R., Renier, C., Hitte, C., André, C., Fraser, C., & Galibert, F. (2003). Comparison of the canine and human olfactory receptor gene repertoires. *Genome Biology*, 4(12), R80. https://doi.org/10.1186/gb-2003-4-12-r80

SLEEP

Kinsman, R., Owczarczak-Garstecka, S., Casey, R., Knowles, T., Tasker, S., Woodward, J., … & Murray, J. (2020). Sleep duration and behaviours: A descriptive analysis of a cohort of dogs up to 12 months of age. *Animals*, 10(7), 1172.

SOCIALISATION

Wormald, D., Lawrence, A., Carter, G., & Fisher, A. (2016). Analysis of correlations between early social exposure and reported aggression in the dog. *Journal of Veterinary Behavior-Clinical Applications and Research*, 15, 31–36.

SOCIAL PAIN

McMillan, F.D. (2016). The psychobiology of social pain: Evidence for a neurocognitive overlap with physical pain and welfare implications for social animals with special attention to the domestic dog (Canis familiaris). *Physiology & Behavior*, 167, 154–171. https://doi.org/10.1016/j.physbeh.2016.09.013

INDEX

Activities, 123–132
dog daycare, 132
enriched environment, 124–128
with other dogs, 128
parallel walks, 128–129
search for treats together, 130
treat search, 123
trips to the dog park, 131
visiting other's houses/gardens, 130, 131
Adolescence; *see also specific entries*
humping, 98
pushing boundaries, 135
risk-taking behaviours, 135
second fear period, 133
Adopting from shelter, 12–13
born in rescues, 12
mixed breeds, 12
rescuing from shelter, 13
separated early, 12
Adult mentors, 106
Alloparenting, x
Anxiety, 12
Attachment-related problems, 12

Barking, 96–97
managing learned barking, 97
reasons for, 96
separation anxiety, 96
work on social skills, 96–97
Brain development, 136
Breeds, 2–3
brachycephalic breeds, 3
extreme features, 3
fluffy poodle crosses, 2
groups and characteristics, 6–9
research, 10
temperaments and inclinations, 3
Breed selection, 5–9
known by sort of work, 5–9
research considerations, 10

Calming signals, 4, 103, 105, 108
Calming Signs: On Talking Terms with Dogs (Turid), 107
Calmness, 41, 100
The Canine Kingdom of Scent (Kvam), 55
Caregiving in loco parentis, 28
Case studies
barking Bran, 53
crate usage for toilet training, 94
habituation, 69
Portuguese water dog joined a young family, 116–117
Tirian, Pinscher puppy and development, 62–63
Case studies, Turid's
Cocker spaniel, four months old, 141–142
Dachshund, four months old, 142–143
German shepherd, five months old, 143
Irish terrier, three and a half months old, 144
Labrador retriever, five months old, 142
Rhodesian ridgeback, three months old, 144
Cats and puppies, 120–122
barriers, 121
in case of not accepting each other, 121–122
co-exist, learning to, 120
introductions, 120
ongoing management, 121
scent, 120

Children and puppies, 109–117
cohabiting, preparing for, 118
dispatches from a dog trainer, 109
enrichment toys for dogs, 111
getting prepared, 111
jealousy, 110
list of rules for other children, 115
making a snuffle mat, 113–114
nipping, 109–110
setting up searches, 112, 115
sleep, 110
things to avoid upsetting the puppy, 111
toys, 110
Choices, 59–63
in bed, 60–61
in food, 60
freedom of choice, 60
in interactions, 61
in movement, 60
and self-determination, 59
in walk, 60
Choosing the right puppy, 10–12
confident and curious, 12
incidence of behavioural disorders, 12
reputable breeder, 10
research on breeders and practices, 10
undesirable characteristics, 11
warning signs, 11
Cleaning product, 14
Comfort, 41; *see also* Three Cs: company, comfort, calmness
Communication, 5, 103–109
adult mentors, 106
calming signals, 103, 108
canine body language, 103–106
human body language, 107–109
observing calming signs, 109
responding to calming signals, 107
spending time with adult dogs, 105
variety of situations, 105, 108
Company, 41; *see also* Three Cs: company, comfort, calmness
Cosy beds, 13
Crates, definition and disadvantages, 17
Creating sense of security and predictability, 48
Custom made mats, 15

Dachshunds, 21
Daycare
arrangement of a babysitter, 132
clear structures and resting time, 132
clear structures and time for resting, 132
well-run daycare, features, 132
Diary, 18
Diet; *see also* Food
food to be added and to be avoided to, 48
fresh food to, 48, 50–51
varied, 48
Dog bags, 14

Eating, common questions on
constantly hungry, 51
faeces-eating (coprophagia), 52
getting picky, 51
leftover food, 52
loose stools, 52
not eating immediately, 51
E-collar, *see* Shock collars
Elimination, 54
dog flap, 54
free access to toileting area, 54
Enriched environment
freedom to interact with environment, 127
household items, 127
new things to investigate., 125
Enrichment activities involving food, *see* Food
Enzyme-removing cleaning products, 14

Favourite things, 19
Fear periods, 25; *see also* Second fear period
Feeding; *see also* Food
guidelines, 47
options, 49

First few weeks, 30–35
developing proprioception, 32
investigating, 32–34
one new 'thing' a day, 30
First trip to the vet, 38
Food, 46–53
amount of, 47
to be added and to be avoided to diet, 48
common questions, 51–53
creating sense of security and predictability, 48
enrichment activities involving food, 46
feeding guidelines, 47
feeding options, 49
frequent meals for puppies, 46
fresh food to diet, 48, 50–51
frustration, 46
handy food resources, 50
kibble or tinned food, 49
raw food, 50
regurgitation, 49
stress-free experience, 47
varied diet, 48
websites, 49
wet food, 49
Food-based chews, 15
Frustration, 46

Good habits, forming, 70–76
to accept being groomed/to stand still for *saumfaring*, 74
to eliminate outside (*see* Toilet training)
loose-lead walking, 71
not to jump on people, 70
to pay attention to the clicking/smacking sound, 70
relaxed showing of palm of the hand, 74
to relax when we sit down, 71–72
to respect the hand signal, 74
to return when called, 75
universal hand signal, 74
to walk on a loose leash, 71

Habits that puppy will not feel good; *see also* Good habits, forming
going into a crate, 75
making eye contact on command, 76
sit-stay, 75–80
walking to heel on your left side, 76
Habituation, 65
Hand signal, 74, 100
Handy food resources, *see* Food
Harness, 16
Humping, 97–98
excitement and stress, 97
paying less attention, 97
Hunting, 3
high instincts, managing tips, 4

Independence
desire to move away and investigate, 134
opportunities for choices and investigation, 135
well-run social walk, 135
Inherent behaviours of all breeds, 3–5
communication, 5
hunting, 3
scavenging, 4
Inter-dog issues, 12
Introducing puppy to new life, 25–30
arriving home, 27
common issues, 30
first night, 27
frequent urination, 26
ready access to food and water, 27
sleeping, 27–29
suitable chew, 26
toileting accidents, 25
travel sickness, 25–26
Itch or a calming sign, 104

Jumping up, 94–97
avoid by turning back and looking away, 95
greeting, 94
at strangers, 95–96

Kibble or tinned food, 49

Lead, 16
Learning life skills
- choosing puppy class, 76–78
- forming good habits, 70–76
- punishment, 78–80
- socialisation and habituation, 65–69

Local breed traditions, 5

Meeting puppy's needs
- common sleep issues, 40–41
- encouraging sleep during the day, 40
- practical sleep solutions, 38–39
- safety, 37
- sleep, 37–38
- three Cs: company, comfort, and calmness, 40–41

Mental stimulation, 55–57
- food based activities, 57
- olfactory enrichment, 56–57
- outlets for using noses, 55
- sniffing, benefits, 56–57
- social interactions, 57

Movement, 42–44
- actually walking, 42
- moving freely around house and garden, 43
- opportunity to explore, 43
- physical challenges, 43–44
- playing fetch, negative effects, 44
- safe places where puppy can run free, 42–43
- slow, deliberate movements, 42

Name tag and collar, 15
Natural chews, 15
Neutering, 136–138
- benefit of neutering, 137
- clear medical indications, 136
- points to keep in mind during decision, 137
- risks of neutering, 137
- 'wait and see' approach, 136

Newspapers or puppy pads for toileting accidents, 14; *see also* Toileting accidents
Nipping, 85–89
- case study, 88–89
- contact seeking, 87
- food-based chew, 86
- frustration, 87
- move away, 88–89
- new environment, 87
- over-stimulation, 87
- pre-empt the behaviour, 88
- preventing, 88
- sleep, 86
- teething, 85
- tiredness, 86
- using our body, 88

Other dogs
- activities with, 128
- learning to interact way with, 131
- parallel walks on-lead walks with, 128
- socialisation, 67

Owners, concerns for; *see also specific entries*
- barking, 96–97
- humping, 97–98
- jumping up, 94–97
- managing separation, 98–101
- puppy chewing on cushion, 86
- toilet training, 89–94
- toothy timeline, 86

Pain; *see also* Punishment
- and discomforting activities, 58–59
- punishment, threat of, 37

Parallel walks, 128–129
- on-lead walks with other dogs, 128
- organising walks, 128
- rest together after short walk, 129

Peed-on carpet, home solutions, 14
Polyphasic sleepers, 37
Positive behaviour change, key points for, 83

Practical sleep solutions, 38–39
    on a dog bed in your room, 39
    on a raised bed beside your bed, 39
    with you in/on the bed, 39
Prefrontal cortex maturation, *see* Brain development
Preparing for puppy, 1–2
    consideration, 1–2
    joy and companionship, 2
Preparing home for puppy, 18–21
    ramp, on to and off higher surfaces, 20
    safety of the items, 19
    slippery floors/stairs and sofas, 20–21
    toileting precautions, 19
    victims of puppy chewing, 19
Punishment, 78–80
    aggressive reaction to confrontational training methods, 79
    'fight or flight' (stress) response, 80
    impact of, 17
    issues, 79
    long-term effect on the dog's behaviour, 79
    reward-based and punishment-based training methods, 80
    risk of an aggressive response to punitive training, 79
Puppy class, choosing, 76–78
    ethos, 77
    mental stimulation, 78
    methods, 78
    recommendations, 78
    scent work, 78
    size, 78
    social groups for puppies, 77
    trainer, 77
Pushing boundaries, 135
    dogs with less secure attachment, 135
    ignore requests, 135

Raised dog bed, 40
Raw food, 50
Regurgitation, 49
Relationships, managing
    communication, 103–109
    with other pets in the household, 118–122
    puppies and children, 109–117
Relationships with other pets in household, 118–122
    allowing to do some teaching, 119
    cats and puppies, 120–122
    downtime to adult dog, 119
    first meeting, 118
    to have plenty of everything, 120
    olfactory introductions, 118
    puppies and adult dogs, 118
    puppy licence, 118
Research on mammalian brains, 28
Risk-taking behaviours, 135

Safety, creating a feeling of, 37
Saumfaring, 35–36
    quick physical check for changes or differences, 35
    touch and stroke, 36
Scavenging, 4
Scout's diary
    day 1, 26
    day 2, 30
    day 5, 31
    day 7, 51
    day 13, 57
    day 18, 87
    day 22, 99
    day 25, 93
    day 32, 55
    night 1, 27
    night 2, 30
    seven months old, 134
Scout update, eight months old
    alone-time, 147
    eating, 147
    exercise, 147
    fear periods and puberty, 148
    sleeping, 147

socialisation and habituation, 148
summary, 148
Searching, 4; *see also* Scavenging
Searching for treats together
spread out for plenty of space, 130
Second fear period, 133–134
during adolescence, 133
support and patience, 133
Self-defence, 4
Separation, managing, 98–101
to avoid separation anxiety, 98
destructive behaviour, 98
house soiling, 98
limitations, 101
minimising stress, 100
psychogenic licking, 98
reduce the lonely time, 101
separation anxiety, 98–101
training for separations, 100–101
universal hand signal, 100
vocalisations, 98
Sexually dimorphic behaviours, 135–136
lifting his leg for urinating, 135
Shock collars, 17
Signs of muscle pain, 36
Sitting, by puppies
discomfort, 82
no sitting or puppies choice, reasons, 81–82
sufficient muscle development, 81
Sleep, 27–29, 37–38; *see also* Practical sleep solutions
choosing beds for puppy, 40
with a member of the family, 29
near family members, 38
negative impact, 41
puppy pile, 29
REM sleep, 37
three Cs: company, comfort, calmness, 40–41
Sniffing, benefits, 56–57; *see also* Mental stimulation
Socialisation, 65–69
clothes and accessories, 67
critical socialisation period, 66
definition, 65
habituation, 65
humans, 67
movement, sound, and reflection, 69
other animals, 68
other dogs, 67
things to consider, 66
things to introduce, 67–68
time to, 66
traffic, 67
Social pain, separation from their social group, 17; *see also* Pain
Social sleeping, 37–38; *see also* Sleep
Stress, 28
not in a position to learn anything, 83
and unable to focus on task, 59
Stress-free experience, 47
Suitable chew, 26

Taking puppy home, 23–25; *see also* Things to be ready
eight weeks old, 23
first fear period in company of his mum, 23, 24
making separation easier, 25
Teaching by mother and siblings, 23
Things to be avoided, 17–18
choke/check/prong collars, 17
crates, 17
extendable leads, 18
head halters, 18
shock collars, 17
Things to be ready, 13–16
bowls, food and water, 14–15
cosy beds, 13
dog bags, 14
food-based chews, 15
good cleaning product, 14
harness, 16

lead, 16
name tag and collar, 15
newspapers or puppy pads for toileting accidents, 14
Three Cs: company, comfort, calmness, 40–41
Toileting accidents, 25
Toilet training, 89–94
cleaning with enzyme-removing products, 91
common issues, 92–93
crates, using, 92
factors affecting, 90
not to tell the dog off for accidents, 91
regular outside visit, 90
reward with tasty treat, 91
scatter-feeding, 91
Training
in canine nutrition, 51
problems, 57
Travel sickness, 25–26
Treat search, 124
Trips to dog park
learning to interact way with other dogs, 131
potential issues with dog parks, 131
risk of physical injury, 131
Troubleshooting, 45–46, 92–93
being restless after walk, 45
biting the lead on walks, 46
periods of excitement, 45
picks things up on walks, 46
puppies not wanting to walk, 45
regression, 93
toilet problems, 92–93

Varied diet, 48
Vet, first trip to, 38
Visiting each other's houses/gardens, 130

Water, constant access to, 53
Websites, 49
Wet food, 49